Home Automation with Intel Galileo

Create thrilling and intricate home automation projects using Intel Galileo

Onur Dundar

PUBLISHING

BIRMINGHAM - MUMBAI

Home Automation with Intel Galileo

First published: March 2015

Production reference: 1240315

Published by Packt Publishing Ltd.
Livery Place
35 Livery Street
Birmingham B3 2PB, UK.

ISBN 978-1-78528-577-6

www.packtpub.com

Credits

Author
  Onur Dundar

Reviewers
  Arun M Kumar

  Mohammed Sharoon

Commissioning Editor
  Edward Bowkett

Acquisition Editor
  Reshma Raman

Content Development Editor
  Vaibhav Pawar

Technical Editors
  Abhishek Kotian

  Prajakta Mhatre

  Faisal Siddiqui

  Anushree Arun Tendulkar

Copy Editors
  Hiral Bhat

  Dipti Kapadia

  Deepa Nambiar

Project Coordinator
  Nidhi Joshi

Proofreaders
  Stephen Copestake

  Safis Editing

  Maria Gould

  Kevin McGowan

Indexer
  Tejal Soni

Graphics
  Sheetal Aute

  Valentina D'silva

Production Coordinator
  Aparna Bhagat

Cover Work
  Aparna Bhagat

About the Author

Onur Dundar is a software engineer who graduated from the Computer Engineering Department of Boğaziçi University. He started working on embedded systems while he was at the university, and in his senior year project, he worked with wireless sensor networks for security and health monitoring.

Onur started his career at JPMorgan Chase & Co. Then, he worked at Intel Corporation for 4 years as a software application engineer. He has specialized in embedded Linux and mostly worked on IPTV platforms, Android, as well as IoT platforms, such as Intel Galileo and Edison, and has developed applications for them. He has also helped software developers enhance their applications on these platforms. He has attended many conferences and developer events to introduce Intel Galileo and Edison. He developed the first application and wrote the public collaterals of Intel Edison when he worked at Intel Corporation.

Onur is continuing his career at Invent Analytics, implementing algorithms for data analytics.

Acknowledgements

I want to thank each member of my family because without their support, I wouldn't have been able to follow my dream to study computer engineering at Boğaziçi University. I had the chance to meet brilliant people and engage in continuous learning at Boğaziçi University.

I want to thank each member of the Computer Engineering Department, starting with Prof. Alper Sen, Prof. Cem Ersoy, and Dr. Hande Alemdar for mentoring me with great projects and courses. I took my tiny steps in engineering with their expertise.

I would like to thank Prof. Arda Yurdakul, who provided me with a great deal of information about embedded systems and engineering. Without her teachings, experience, and knowledge of embedded engineering as well as her extensive support, I wouldn't have been able to make a career in embedded systems and enjoy my job.

Oktay Ozgun, who is the brightest, is one of the most respectful people I've ever met and the best person to have as a supervisor. He always created the opportunity for me to work on great projects and shared his vision, knowledge, and experience. If I hadn't worked with Oktay Ozgun, I wouldn't have been as passionate about software, science, and engineering as I am today.

Brendan Le Foll is one of the best people in the world to learn the best practices of Linux. He always shared his expertise and expanded my knowledge about Linux. If I hadn't met Brendan, I wouldn't have had this much knowledge about Linux.

Last but not least, I want to thank Steve Cutler, Andrew John, Marcel Wagner, Rami Radi, Peter Rohr, and Alex Klimovitski for their extensive support and for always creating the best environment while working on great projects during my time at Intel Corporation.

I want to thank all of them; if I had not met these great people and worked with them, I wouldn't have been able to have the knowledge and experience to write this book.

Finally, I want to thank my beloved Canan Soysal for her extensive support while authoring this book in my busy schedule.

About the Reviewers

Arun M Kumar is a young tech guy, who borderlines somewhere between a geek and a nerd. He reads tech articles to relax and is more inclined to call himself a generalist rather than a specialist. Arun has an unexplained phobia toward orange-flavored candies and still prefers C as the de facto language of the computing world. Arun loves to work on microcontrollers and has an unjustified dislike toward the Arduino environment. Working on open source projects is his first love, and he also loves to spend time travelling with friends. Having worked in automotive, power, and cloud computing domains, Arun finds pleasure in having the same random set of friends and experiences. His aim in life is to travel, explore, and spread smiles wherever he goes while still figuring out what to try next.

This being the first book I reviewed, I would like to thank Nidhi Joshi and others from Packt Publishing for providing this opportunity and for putting up with my busy schedule during the review process. I would also like to thank CDAC-ACTS, Pune, and my classmates there for giving me the skills that made me what I am today. I would also like to thank Prachee Sonchal and the rest of the HR team in my organization for supporting me in various ways.

Mohammed Sharoon is a final year electronics engineering student at National Institute of Technology, Calicut, with experience in implementing simple embedded projects. He hopes to develop this interest further by gaining a better perspective of the numerous domains of electronics in order to develop breakthroughs that will make our lives easier.

I would like to thank my best friend for providing me the motivation to push myself and achieve more, without which nothing would have been possible.

www.PacktPub.com

Support files, eBooks, discount offers, and more

For support files and downloads related to your book, please visit www.PacktPub.com.

Did you know that Packt offers eBook versions of every book published, with PDF and ePub files available? You can upgrade to the eBook version at www.PacktPub.com and as a print book customer, you are entitled to a discount on the eBook copy. Get in touch with us at service@packtpub.com for more details.

At www.PacktPub.com, you can also read a collection of free technical articles, sign up for a range of free newsletters and receive exclusive discounts and offers on Packt books and eBooks.

https://www2.packtpub.com/books/subscription/packtlib

Do you need instant solutions to your IT questions? PacktLib is Packt's online digital book library. Here, you can search, access, and read Packt's entire library of books.

Why subscribe?

- Fully searchable across every book published by Packt
- Copy and paste, print, and bookmark content
- On demand and accessible via a web browser

Free access for Packt account holders

If you have an account with Packt at www.PacktPub.com, you can use this to access PacktLib today and view 9 entirely free books. Simply use your login credentials for immediate access.

Table of Contents

Preface

All of us have heard about the Internet of Things. Every new device is being developed around this new concept; all of them being connected. Our residences can't be outside of this world. We can design and create new use cases to connect every device we own at home to manage them through the Internet.

In this book, we have used the Intel Galileo development board to show the various methods to make a connected home using open source software and the Linux operating system to develop applications. We have tried to cover all common devices and sensors with sample applications developed with the C programming language, hoping to be an inspiration to help you make greater home automation applications.

We have tried to end the book by merging all the sample applications into one home automation application to manage devices remotely to make a connected home and easily automate and manage this with a smartphone, tablet, or a PC.

What this book covers

Chapter 1, *Getting Started with Intel Galileo*, introduces the Intel Galileo development board and presents step-by-step instructions for you to set up a development environment to get started with application development.

Chapter 2, *Getting Started with Home Automation Applications*, explains home automation concepts and introduces existing technologies and open source projects related to home automation.

Chapter 3, *Energy Management with Environmental and Electrical Sensors*, introduces sensors and devices for use in a home automation application. This chapter takes you inside the energy management with temperature sensors and power meters.

Chapter 4, Energy Management with Light Sensors and Extending Use Cases, introduces new sensors, which you can use to manage the lighting of your home with sample applications.

Chapter 5, Home Monitoring with Common Security Sensors, introduces new sensors and devices you can use to add security in your home. This chapter presents example applications to instruct you on how to use these sensors.

Chapter 6, Home Surveillance and Extending Security Use Cases, gets you inside the surveillance world with the help of a camera. This chapter tells you how you can use a camera with Intel Galileo and include it in your home automation application.

Chapter 7, Building Applications and Customizing Linux for Home Automation, introduces the basics of the Yocto project and how you can customize Linux to make Intel Galileo run your application and make it ready to serve as a home automation hub.

Chapter 8, Extending Use Cases, introduces other technologies available that can be used with Intel Galileo. We provide Node.js and Android application samples to extend our home automation application with a better user experience and user interface.

What you need for this book

For this book, you will need following hardware peripherals and software:

- A computer running Ubuntu 12.04 or other Linux distribution (a virtual machine with Ubuntu 12.04 can be used as well depending on your preference).
- An Intel Galileo Generation 2 development board.
- A microSD card.
- An Ethernet cable to connect Intel Galileo to your local network or router (if you have a PCI-e Wi-Fi card, you can use it to connect your Intel Galileo to your Wi-Fi router).
- A USB to TTL serial cable (3.5 mm jack to USB serial cable is needed if you already have Intel Galileo Generation 1); the suggested cable is TTL-232R-3V3.
- A micro USB to USB 2.0 cable.
- Sensirion SHT11 or SHT15 temperature sensor.

The following Z-Wave devices need to be selected according to your region. Each region (US, EU, and so on) has a different Z-Wave frequency. In this book, we have used EU devices.

- A Z-Wave Aeon Labs Z-Stick S2 USB controller.
- A Z-Wave Fibaro Wall plug.
- Z-Wave Philio Multisensor — door / PIR / light / temperature.
- A Z-Wave Everspring screw-in module.
- A Z-Wave Everspring flood sensor.
- MQ-9 gas sensor.
- D-Link DCS-930L IP camera (any other IP cam or USB cam can also be used with its own configurations).
- An Android smartphone (it is optional; in the last chapter, we develop a sample Android application. You can also use an Android virtual device. If you are already able to develop for other mobile devices, iOS, or the Windows phone, you can develop a similar sample application for them.)

Who this book is for

This book is aimed at developers, hobbyists, and makers who have some experience with Linux, C, and Arduino programming and want to explore opportunities in the home automation world. The book also includes some introductory examples and practices for people who are interested in starting software development for Intel Galileo, similar devices, and embedded Linux.

Conventions

In this book, you will find a number of text styles that distinguish between different kinds of information. Here are some examples of these styles and an explanation of their meaning.

Code words in text are shown as follows: "The following piece of code shows two functions `setGPIO(int number, int direction)` and `getGPIO(int number)`."

A block of code is set as follows:

```
float readTemperature(){
   int register_address = 0x4600;
   return getTemperature(register_address);
}
```

When we wish to draw your attention to a particular part of a code block, the relevant lines or items are set in bold:

```
float readTemperature(){
   int register_address = 0x4600;
return getTemperature(register_address);
}
```

Any command-line input or output is written as follows:

```
$ cp /usr/src/asterisk-addons/configs/cdr_mysql.conf.sample
   /etc/asterisk/cdr_mysql.conf
```

New terms and **important words** are shown in bold. Words that you see on the screen, for example, in menus or dialog boxes, appear in the text like this: "Now we can click on the **Fibaro Wall Plug** item on the list to switch it on or off from our smartphone."

URLs in the text are shown as follows: https://www.packtpub.com

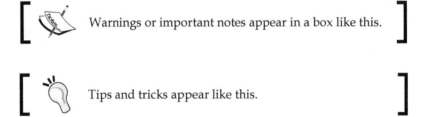

Warnings or important notes appear in a box like this.

Tips and tricks appear like this.

Reader feedback

Feedback from our readers is always welcome. Let us know what you think about this book—what you liked or disliked. Reader feedback is important for us as it helps us develop titles that you will really get the most out of.

To send us general feedback, simply e-mail feedback@packtpub.com, and mention the book's title in the subject of your message.

If there is a topic that you have expertise in and you are interested in either writing or contributing to a book, see our author guide at www.packtpub.com/authors.

Customer support

Now that you are the proud owner of a Packt book, we have a number of things to help you to get the most from your purchase.

Downloading the example code

You can download the example code files from your account at http://www.packtpub.com for all the Packt Publishing books you have purchased. If you purchased this book elsewhere, you can visit http://www.packtpub.com/support and register to have the files e-mailed directly to you.

Downloading the color images of this book

We also provide you with a PDF file that has color images of the screenshots/diagrams used in this book. The color images will help you better understand the changes in the output. You can download this file from: https://www.packtpub.com/sites/default/files/downloads/5776OS_ColoredImages.pdf.

Errata

Although we have taken every care to ensure the accuracy of our content, mistakes do happen. If you find a mistake in one of our books—maybe a mistake in the text or the code—we would be grateful if you could report this to us. By doing so, you can save other readers from frustration and help us improve subsequent versions of this book. If you find any errata, please report them by visiting http://www.packtpub.com/submit-errata, selecting your book, clicking on the **Errata Submission Form** link, and entering the details of your errata. Once your errata are verified, your submission will be accepted and the errata will be uploaded to our website or added to any list of existing errata under the Errata section of that title.

To view the previously submitted errata, go to https://www.packtpub.com/books/content/support and enter the name of the book in the search field. The required information will appear under the **Errata** section.

Piracy

Piracy of copyrighted material on the Internet is an ongoing problem across all media. At Packt, we take the protection of our copyright and licenses very seriously. If you come across any illegal copies of our works in any form on the Internet, please provide us with the location address or website name immediately so that we can pursue a remedy.

Please contact us at copyright@packtpub.com with a link to the suspected pirated material.

We appreciate your help in protecting our authors and our ability to bring you valuable content.

Questions

If you have a problem with any aspect of this book, you can contact us at questions@packtpub.com, and we will do our best to address the problem.

1

Getting Started with Intel Galileo

This book is about developing home automation examples using the Intel Galileo development board along with the existing home automation sensors and devices. In the book, a good review of Intel Galileo will be provided, which will teach you to develop native C/C++ applications for Intel Galileo.

After a good introduction to Intel Galileo, we will review home automation's history, concepts, technology, and current trends. When we have an understanding of home automation and the supporting technologies, we will develop some examples on two main concepts of home automation: energy management and security.

We will build some examples under energy management using electrical switches, light bulbs and switches, as well as temperature sensors. For security, we will use motion, water leak sensors, and a camera to create some examples. For all the examples, we will develop simple applications with C and C++.

Finally, when we are done building good and working examples, we will work on supporting software and technologies to create more user friendly home automation software.

In this chapter, we will take a look at the Intel Galileo development board, which will be the device that we will use to build all our applications; also, we will configure our host PC environment for software development.

The following are the prerequisites for this chapter:

- A Linux PC for development purposes. All our work has been done on an Ubuntu 12.04 host computer, for this chapter and others as well. (If you use newer versions of Ubuntu, you might encounter problems with some things in this chapter.)

- An Intel Galileo (Gen 2) development board with its power adapter.

- A USB-to-TTL serial UART converter cable; the suggested cable is TTL-232R-3V3 to connect to the Intel Galileo Gen 2 board and your host system. You can see an example of a USB-to-TTL serial UART cable at `http://www.amazon.com/GearMo%C2%AE-3-3v-Header-like-TTL-232R-3V3/dp/B004LBX02A`. If you are going to use Intel Galileo Gen 1, you will need a 3.5 mm jack-to-UART cable. You can see the mentioned cable at `http://www.amazon.com/Intel-Galileo-Gen-Serial-cable/dp/B000170JKY/`.

- An Ethernet cable connected to your modem or switch in order to connect Intel Galileo to the local network of your workplace.

- A microSD card. Intel Galileo supports microSD cards up to 32 GB storage.

Introducing Intel Galileo

The Intel Galileo board is the first in a line of **Arduino**-certified development boards based on Intel x86 architecture. It is designed to be hardware and software pin-compatible with Arduino shields designed for the UNOR3.

> Arduino is an open source physical computing platform based on a simple microcontroller board, and it is a development environment for writing software for the board. Arduino can be used to develop interactive objects, by taking inputs from a variety of switches or sensors and controlling a variety of lights, motors, and other physical outputs.

The Intel Galileo board is based on the Intel Quark X1000 SoC, a 32-bit Intel Pentium processor-class system on a chip (SoC). In addition to Arduino compatible I/O pins, Intel Galileo inherited mini PCI Express slots, a 10/100 Mbps Ethernet RJ45 port, USB 2.0 host, and client I/O ports from the PC world.

> The Intel Galileo Gen 1 USB host is a micro USB slot. In order to use a generation 1 USB host with USB 2.0 cables, you will need an OTG (On-the-go) cable. You can see an example cable at `http://www.amazon.com/Cable-Matters-2-Pack-Micro-USB-Adapter/dp/B00GM00Z40`.

Another good feature of the Intel Galileo board is that it has open source hardware designed together with its software. Hardware design schematics and the **bill of materials (BOM)** are distributed on the Intel website. Intel Galileo runs on a custom embedded Linux operating system, and its firmware, bootloader, as well as kernel source code can be downloaded from `https://downloadcenter.intel.com/Detail_Desc.aspx?DwnldID=23171`.

> Another helpful URL to identify, locate, and ask questions about the latest changes in the software and hardware is the open source community at `https://communities.intel.com/community/makers`.

Intel delivered two versions of the Intel Galileo development board called Gen 1 and Gen 2. At the moment, only Gen 2 versions are available. There are some hardware changes in Gen 2, as compared to Gen 1. You can see both versions in the following image:

The first board (on the left-hand side) is the Intel Galileo Gen 1 version and the second one (on the right-hand side) is Intel Galileo Gen 2.

Using Intel Galileo for home automation

As mentioned in the previous section, Intel Galileo supports various sets of I/O peripherals. Arduino sensor shields and USB and mini PCI-E devices can be used to develop and create applications. Intel Galileo can be expanded with the help of I/O peripherals, so we can manage the sensors needed to automate our home.

When we take a look at the existing home automation modules in the market, we can see that preconfigured hubs or gateways manage these modules to automate homes. A hub or a gateway is programmed to send and receive data to/from home automation devices. Similarly, with the help of a Linux operating system running on Intel Galileo and the support of multiple I/O ports on the board, we will be able to manage home automation devices.

We will implement new applications or will port existing Linux applications to connect home automation devices. Connecting to the devices will enable us to collect data as well as receive and send commands to these devices. Being able to send and receive commands to and from these devices will make Intel Galileo a gateway or a hub for home automation.

It is also possible to develop simple home automation devices with the help of the existing sensors. Pinout helps us to connect sensors on the board and read/write data to sensors and come up with a device.

Finally, the power of open source and Linux on Intel Galileo will enable you to reuse the developed libraries for your projects. It can also be used to run existing open source projects on technologies such as Node.js and Python on the board together with our C application. This will help you to add more features and extend the board's capability, for example, serving a web user interface easily from Intel Galileo with Node.js.

Intel Galileo – hardware specifications

The Intel Galileo board is an open source hardware design. The schematics, Cadence Allegro board files, and BOM can be downloaded from the Intel Galileo web page.

In this section, we will just take a look at some key hardware features for feature references to understand the hardware capability of Intel Galileo in order to make better decisions on software design.

Intel Galileo is an embedded system with the required RAM and flash storages included on the board to boot it and run without any additional hardware.

The following table shows the features of Intel Galileo:

Processor features	• 1 Core 32-bit Intel Pentium processor-compatible ISA Intel Quark SoC X1000 • 400 MHz • 16 KB L1 Cache • 512 KB SRAM • Integrated **real-time clock (RTC)**
Storage	• 8 MB NOR Flash for firmware and bootloader • 256 MB DDR3; 800 MT/s • SD card, up to 32 GB • 8 KB EEPROM
Power	• 7 V to 15 V • **Power over Ethernet (PoE)** requires you to install the PoE module
Ports and connectors	• USB 2.0 host (standard type A), client (micro USB type B) • RJ45 Ethernet • 10-pin JTAG for debugging • 6-pin UART • 6-pin ICSP • 1 mini-PCI Express slot • 1 SDIO
Arduino compatible headers	• 20 digital I/O pins • 6 analog inputs • 6 PWMs with 12-bit resolution • 1 SPI master • 2 UARTs (one shared with the console UART) • 1 I2C master

Intel Galileo – software specifications

Intel delivers prebuilt images and binaries along with its **board support package (BSP)** to download the source code and build all related software with your development system.

The running operating system on Intel Galileo is Linux; sometimes, it is called **Yocto Linux** because of the Linux filesystem, cross-compiled toolchain, and kernel images created by the Yocto Project's build mechanism.

 The *Yocto* Project is an open source collaboration project that provides templates, tools, and methods to help you create custom Linux-based systems for embedded products, regardless of the hardware architecture.

The following diagram shows the layers of the Intel Galileo development board:

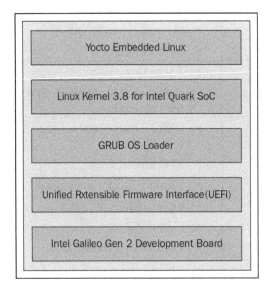

Intel Galileo is an embedded Linux product; this means you need to compile your software on your development machine with the help of a cross-compiled toolchain or **software development kit (SDK)**.

A cross-compiled toolchain/SDK can be created using the Yocto project; we will go over the instructions in the following sections. The toolchain includes the necessary compiler and linker for Intel Galileo to compile and build C/C++ applications for the Intel Galileo board. The binary created on your host with the Intel Galileo SDK will not work on the host machine since it is created for a different architecture.

With the help of the C/C++ APIs and libraries provided with the Intel Galileo SDK, you can build any C/C++ native application for Intel Galileo as well as port any existing native application (without a graphical user interface) to run on Intel Galileo.

Intel Galileo doesn't have a graphical processor unit. You can still use OpenCV-like libraries, but the performance of matrix operations is so poor on CPU compared to systems with GPU that it is not wise to perform complex image processing on Intel Galileo.

Connecting and booting Intel Galileo

We can now proceed to power up Intel Galileo and connect it to its terminal.

Before going forward with the board connection, you need to install a modem control program to your host system in order to connect Intel Galileo from its UART interface with **minicom**.

 Minicom is a text-based modem control and terminal emulation program for Unix-like operating systems. If you are not comfortable with text-based applications, you can use graphical serial terminals such as **CuteCom** or **GtkTerm**.

To start with Intel Galileo, perform the following steps:

1. Install minicom:

    ```
    $ sudo apt-get install minicom
    ```

     Downloading the example code

    You can download the example code files from your account at http://www.packtpub.com for all the Packt Publishing books you have purchased. If you purchased this book elsewhere, you can visit http://www.packtpub.com/support and register to have the files e-mailed directly to you.

2. Attach the USB of your 6-pin TTL cable and start minicom for the first time with the -s option:

    ```
    $ sudo minicom -s
    ```

3. Before going into the setup details, check the device is connected to your host. In our case, the serial device is /dev/ttyUSB0 on our host system. You can check it from your host's device messages (dmesg) to see the connected USB.

4. When you start minicom with the `-s` option, it will prompt you. From minicom's **Configuration** menu, select **Serial port setup** to set the values, as follows:

```
+----------------------------------------------------------------+
| A -    Serial Device       : /dev/ttyUSB0                      |
| B - Lockfile Location      : /var/lock                         |
| C -    Callin Program      :                                   |
| D -   Callout Program      :                                   |
| E -     Bps/Par/Bits       : 115200 8N1                        |
| F - Hardware Flow Control  : No                                |
| G - Software Flow Control  : No                                |
|                                                                |
|    Change which setting?                                       |
+----------------------------------------------------------------+
```

5. After setting up the serial device, select **Exit** to go to the terminal. This will prompt you with the booting sequence and launch the Linux console when the Intel Galileo serial device is connected and powered up.

6. Next, complete connections on Intel Galileo. Connect the TTL-232R cable to your Intel Galileo board's UART pins. UART pins are just next to the Ethernet port.

   Make sure that you have connected the cables correctly. The black-colored cable on TTL is the ground connection. It is written on TTL pins which one is ground on Intel Galileo.

7. We are ready to power up Intel Galileo. After you plug the power cable into the board, you will see the Intel Galileo board's boot sequence on the terminal. When the booting process is completed, it will prompt you to log in; log in with the `root` user, where no password is needed.

8. The final prompt will be as follows; we are in the Intel Galileo Linux console, where you can just use basic Linux commands that already exist on the board to discover the Intel Galileo filesystem:

```
Poky 9.0.2 (Yocto Project 1.4 Reference Distro) 1.4.2
  clanton

clanton login: root
root@clanton:~#
```

Your board will now look like the following image:

Connecting to Intel Galileo via Telnet

If you have connected Intel Galileo to a local network with an Ethernet cable, you can use Telnet to connect it without using a serial connection, after performing some simple steps:

1. Run the following commands on the Intel Galileo terminal:

   ```
   root@clanton:~# ifup eth0
   root@clanton:~# ifconfig
   root@clanton:~# telnetd
   ```

2. The `ifup` command brings the Ethernet interface up, and the second command starts the Telnet daemon. You can check the assigned IP address with the `ifconfig` command.

3. From your host system, run the following command with your Intel Galileo board's IP address to start a Telnet session with Intel Galileo:

   ```
   $ telnet 192.168.2.168
   ```

Building a Linux image for Intel Galileo with the Yocto Project

We went through the software specifications of Intel Galileo in the previous section and booted Intel Galileo with the Linux image on its SPI flash.

In this section, we are going to cover how to build a customized image for Intel Galileo with some additional software packages using the Yocto Project, and we will boot Intel Galileo from the microSD card with our new Linux image.

>
> The Yocto Project is an open source project that helps embedded Linux developers by providing a set of tools such as *Poky* to ease the customization of Linux filesystems, building kernel images. This project uses a folder structure to store the metadata of the build information of individual software projects. Each software application or library has a metadata file called `recipes` with the .bb and .bbclass files. A quick start guide for developers is available at `http://www.yoctoproject.org/docs/current/yocto-project-qs/yocto-project-qs.html`, where you can get the basics of the Yocto Project.

You will also learn some basics of the Yocto Project to build and customize Linux filesystems for Intel Galileo.

>
> There are prebuilt SD card images for Intel Galileo; you can easily get them from the Intel download page at `https://communities.intel.com/docs/DOC-22226`.

Learning about the build process will teach you how to customize Linux for future needs. Before that, the following prerequisites are needed:

1. We need to first download the Intel Quark board support package from `http://downloadcenter.intel.com/confirm.aspx?httpDown=http://downloadmirror.intel.com/23197/eng/Board_Support_Package_Sources_for_Intel_Quark_v1.0.1.7z&Lang=eng&Dwnldid=23197`.

2. Then, we should download BSP patches to fix some of the problems with upstream sources. Download the patches from `https://github.com/01org/Galileo-Runtime/archive/1.0.4.tar.gz`. Patching instructions can be found in the following link: `http://downloadmirror.intel.com/24355/eng/BSP-Patches-and-Build_Instructions.1.0.4.txt`.

3. The next step is to extract the files. As the BSP package is distributed in 7-Zip format, make sure that you have 7-Zip installed on your host:

   ```
   $ sudo apt-get install p7zip
   ```

A good way to deal with the build process and all the mess created is to create a unique folder in our `home` directory, such as `/home/onur/galileo_build`, and do all the build work in that directory. I will refer to our build in the directory as `BUILD_DIR`.

Building Linux filesystems for Intel Galileo

You need to follow these steps to build a Linux filesystem:

1. Put your downloaded files in your build directory and extract them, as follows:

   ```
   $ cd /home/onur/galileo_build
   $ mv
   ~/Downloads/board_support_package_sources_for_intel_quark_v1.0
     .1.7z .
   $ mv ~/Downloads/BSP-Patches-and-
     Build_Instructions.1.0.4.tar.bz2 .
   $ 7z x
     board_support_package_sources_for_intel_quark_v1.0.1.7z
   ```

   BSP includes packages for the layers of software for Intel Galileo. *Grub OS Loader, Linux Filesystem build files for Yocto Project, EDKII (Firmware for Quark), Linux Kernel for Intel Quark, SPI-Flash tools*, and *System Image files* are the packages required by developers to rebuild and reuse. Our focus will be on the `meta-clanton_v1.0.1.tar.gz` file to create the Linux filesystem to boot with the SD card.

    You will see **clanton** in many places in the files you downloaded. It refers to systems with Intel Quark processors.

   The `board_support_package_sources_for_intel_quark_v1.0.1.7z` file includes the following compressed files:

   - `grub-legacy_5775f32a+v1.0.1.tar.gz`
   - `meta-clanton_v1.0.1.tar.gz`
   - `Quark_EDKII_v1.0.1.tar.gz`
   - `quark_linux_v3.8.7+v1.0.1.tar.gz`
   - `spi-flash-tools_v1.0.1.tar.gz`
   - `sysimage_v1.0.1.tar.gz`
   - `sha1sum.txt`

2. The `BSP-Patches-and-Build_Instructions` file includes a folder called `patches`. It has a number of patches to apply the Yocto Project metadata (recipe) `.bb` files:

    ```
    $ tar xvf BSP-Patches-and-Build_Instructions.1.0.4.tar.bz2
    ```

3. Extract metadata for the Intel Galileo development board to build the Linux filesystem. Metadata files will be extracted into the `meta-clanton_v1.0.1` directory:

    ```
    $ tar xvf meta-clanton_v1.0.1.tar.gz
    ```

> It is highly recommended that you apply patches inside the extracted `patches` folder that comes along with the `BSP_Patches_and_Instructions` file. Instructions are stored in the `patches/patches.txt` file. If you don't apply the patches before starting the build process, you are highly likely to get errors.

4. There are a couple of tools that you need to install on your host system to start the building process. Make sure you have installed them:

    ```
    $ sudo apt-get install git diffstat texinfo gawk chrpath file
      build-essential gcc-multilib chrpath
    ```

5. Go to the `meta-clanton_v1.0.1` folder to start the build process:

    ```
    $ cd $BUILD_DIR/meta-clanton_v1.0.1
    ```

    This is where all the metadata is placed for building Linux filesystems. After applying the patches, we can start executing the scripts to start building:

6. The first step is to run the `setup.sh` script in the `meta-clanton_v1.0.1` folder to get the required external sources. It will also create the folder `yocto_build` with the required configuration files to define metadata layers for the Yocto build tool BitBake:

    ```
    $ ./setup.sh
    ```

7. Then, we need to initialize the environment variables and specify the build folder for the output of the build process:

    ```
    $ source poky/oe-init-build-env yocto_build
    ```

> If you've closed your current shell session and started on a new one, for each shell that you open, you need to source environment variables with `oe-init-build-env`.

After you initialize the environment variables, you will be redirected to the yocto_build folder. This is the folder where all the downloaded sources and the output of the build process will be copied.

Now we are ready to start the build process with the Yocto Project command tool **BitBake**.

 BitBake take cares of the entire build process by parsing configuration files in all the layers, the metadata (recipes), classes, and configurations files such as the .bb, .bbclass, and .conf files, respectively.

There is already a metadata file defined to build a full image, which includes many open source projects such as Node.js, OpenCV, and additional kernel modules to support mini PCI-E Wi-Fi cards.

8. Start the build process with the following command:

    ```
    $ bitbake image-full-galileo
    ```

9. An output similar to the following will be seen on your host machine:

    ```
    Loading cache: 100% |#########################################
    ##################################################################
    ##################################################################
    #########################| ETA:  00:00:00
    Loaded 1617 entries from dependency cache.

    Build Configuration:
    BB_VERSION       = "1.18.0"
    BUILD_SYS        = "x86_64-linux"
    NATIVELSBSTRING  = "Ubuntu-12.04"
    TARGET_SYS       = "i586-poky-linux-uclibc"
    MACHINE          = "clanton"
    DISTRO           = "clanton-tiny"
    DISTRO_VERSION   = "1.4.2"
    TUNE_FEATURES    = "m32 i586"
    TARGET_FPU       = ""
    meta
    meta-yocto
    meta-yocto-bsp   =
      "clanton:d734ab491a30078d43dee5440c03acce2d251425"
    meta-intel       =
      "clanton:048def7bae8e3e1a11c91f5071f99bdcf8e6dd16"
    meta-oe          =
      "clanton:13ae5105ee30410136beeae66ec41ee4a8a2e2b0"
    meta-clanton-distro
    meta-clanton-bsp = "<unknown>:<unknown>"
    ```

```
Currently 12 running tasks (179 of 2924):
0: uclibc-initial-
   0.9.33+gitAUTOINC+946799cd0ce0c6c803c9cb173a84f4d607bde350-
   r8.4 do_unpack (pid 32309)
1: binutils-cross-2.23.1-r3 do_unpack (pid 32304)
2: linux-libc-headers-3.8-r0 do_fetch (pid 32307)
3: gcc-cross-initial-4.7.2-r20 do_fetch (pid 32308)
4: libmpc-native-0.8.2-r1 do_compile (pid 32305)
5: python-native-2.7.3-r0.1 do_unpack (pid 32316)
6: uclibc-
   0.9.33+gitAUTOINC+946799cd0ce0c6c803c9cb173a84f4d607bde350-
   r8.4 do_unpack (pid 32310)
7: elfutils-native-0.148-r11 do_compile (pid 32314)
8: file-native-5.13-r0 do_compile (pid 32315)
9: readline-native-6.2-r4 do_configure (pid 32311)
10: openssl-native-1.0.1h-r15.0 do_install (pid 32312)
11: attr-native-2.4.46-r4 do_configure (pid 32313)
```

The build process can take around 2 hours to finish, depending on the processing power of your host machine.

The build process will start by parsing recipes in the board's support package and will fetch the source code, configure, build, and install on the final Linux filesystem image.

If everything goes well and the build process finishes successfully, all the required files will be created in the `$BUILD_DIR/meta_clanton_v1.0.1/yocto_build/tmp/deploy/images` folder to be used for your SD card.

Preparing the SD card to boot

When the build process is successful, you can go ahead and copy the required files onto your microSD card to boot with Intel Galileo.

First, you need a microSD card; Intel Galileo supports SD cards up to 32 GB in capacity. Format your microSD card as FAT32 for first-time use and then copy the following files to your microSD card:

1. Format the SD card as FAT32, and check the assigned device file for the SD card on your system; something such as `/dev/sdd` or `/dev/mmcblk0` should be assigned to it. You can use device messages (the `dmesg` command) to check the assigned device file for the SD card. Run the `dmesg` command before and after you have attached the SD card on your host PC terminal. Then, you can see the assigned device file. In this section, we will use `/dev/sdX` to indicate the SD card's device file:

   ```
   $ sudo mkfs.msdos /dev/sdX
   ```

2. Mount the SD card on your host system:

```
$ sudo mount /dev/sdX /mnt/sdcard
```

3. Copy the following files to your SD card:

```
$ cd $BUILD_DIR/meta-
    clanton_v1.0.1/yocto_build/tmp/deploy/images/
```

```
$ cp image-full-galileo-clanton.ext3 core-image-minimal-
    initramfs-clanton.cpio.gz bzImage grub.efi -t /mnt/sdcard
```

```
$ cp -r boot/ -t /mnt/sdcard
```

 The image-full-galileo-clanton.ext3 file includes the Linux root filesystem. The bzImage file is the Linux kernel image. The core-image-minimal-initramfs-clanton.cpio.gz file is the initial RAM file system. grub.efi is the file, and GRUB/UEFI is the firmware for Intel Galileo. The boot folder includes the GRUB configuration.

4. Unmount the SD card and detach it from your host system:

```
$ sudo unmount /mnt/sdcard
```

5. Insert the microSD card into the slot next to the power port on the Intel Galileo development board.

Connect the serial cable, as shown in the previous section, and power up Intel Galileo. When Intel Galileo is booting from the SD card, an LED starts to blink. This is the LED on which the SD card is writing.

When you are prompted to log in, just like when you booted the SPI image in the previous section, you can log in with the root user and wander around the additional installed packages with the SD image.

An advantage of using an SD card instead of using an SPI flash image is that you will have more storage, so you can install more software.

Created files are also not volatile on an SD card; they will not be removed when you reboot the board. If you are using the Intel Galileo board from the SPI flash image, it will remove all the created files when you reboot the board.

Upgrading firmware on Intel Galileo

While booting from the SD card, you may encounter problems. If the version of firmware on the Intel Galileo board and the BSP version that you build on the Linux filesystem don't match, the board doesn't boot the SD card. You can easily see this as the SD card's LED doesn't blink.

Check whether you formatted the SD card correctly to FAT32 and the files have been copied correctly. If the SD card format and files are not corrupted, it is suggested that you upgrade the firmware on the Intel Galileo board you have, as follows:

1. Download the Arduino IDE for Intel Galileo, for your host computer's architecture, from `https://communities.intel.com/docs/DOC-22226`.

2. Connect Intel Galileo to your PC using a micro USB-to-USB cable. When Intel Galileo is connected to the host device with a micro USB, it defines a new device, `/dev/ttyACM0`. You can check the device messages (`dmesg`) to check whether Intel Galileo is connected correctly.

3. Extract the downloaded file:

    ```
    $ tar xvf arduino-linux64-1.0.4.tgz
    ```

4. Go to the `arduino-linux64-1.0.4.tgz` folder and run the `Arduino IDE` executable with `sudo`:

    ```
    $ cd arduino-linux-1.0.4.tgz
    ```

5. In the menu bar of the Arduino IDE, navigate to the following:
    *   **Tools | Serial Port | /dev/ttyACM0**
    *   **Tools | Board | Intel Galileo (Gen 2)**

6. Check for the firmware upgrade by navigating to **Help | Galileo Firmware Update**.

If the firmware on your device has a lower version, the Arduino IDE will prompt you to upgrade. Click on **OK** to upgrade the firmware and follow the instructions. This should take approximately 5 to 10 minutes. The Arduino IDE will prompt when the upgrade has been successfully installed.

Building the Intel Galileo SDK

We have gone through the steps to get the board's support packages and to build the Linux kernel as well as the filesystem. Since we are going to develop and build applications for Intel Galileo, we need an SDK to compile and build.

The Yocto Project allows you to easily create the SDK for embedded devices, similar to creating a filesystem. We need to run one more command right after we build the filesystem.

All the commands used to build an SDK are similar to the commands used to build a filesystem, but only the last command changes to create a toolchain, as follows:

```
$ bitbake image-full-galileo -c populate_sdk
```

After the successful execution of the populate_sdk command, the SDK installer will be deployed into the $BUILD_DIR/meta_clanton_v1.0.1/yocto_build/tmp/deploy/sdk folder:

```
$ ls tmp/deploy/sdk
clanton-tiny-uclibc-x86_64-i586-toolchain-1.4.2.sh
```

In this example, BitBake created an image for a 64-bit host; if you build on a 32-bit host, BitBake will create a toolchain for a 32-bit architecture.

> It is also possible to create toolchains for other architectures. Add the SDKMACHINE ?= i686 line to the yocto_build/conf/local.conf file and rerun the command. It will create the file clanton-tiny-uclibc-i686-i586-toolchain-1.4.2.sh.

Setting up a development environment for Intel Galileo

In this section, we will go through the two ways of setting up a development environment for Intel Galileo on our host development system. In the previous section, we created the Intel Galileo SDK, and now we will go through how to use it:

1. Go to the deployment folder of your build directory:

    ```
    $ cd $BUILD_DIR/meta-clanton/yocto_build/.tmp/deploy/sdk
    ```

2. Run the SDK installer to install the Intel Galileo toolchain on your host filesystem. By default, the installation directory has been set to the /opt/clanton-tiny/1.4.2 directory; you can install any directory you want. You will get the following terminal output when you run the SDK installer script:

    ```
    $ sudo ./clanton-tiny-uclibc-x86_64-i586-toolchain-1.4.2.sh
    Enter target directory for SDK (default: /opt/clanton-tiny/1.4.2):
    ```

```
You are about to install the SDK to "/opt/clanton-tiny/1.4.2".
Proceed[Y/n]?Y

Extracting SDK...

Setting it up...done

SDK has been successfully set up and is ready to be used.
```

The Intel Galileo SDK is installed on your host filesystem, and your host system is ready to compile and build C/C++ applications for the Intel Galileo development board.

To start development, you can either use basic text editors and compile the source code on your host machine's console, or you can configure an IDE to use the Intel Galileo SDK to develop applications.

Setting up a development environment for Linux

If you don't want to use any IDE for software development, you can just use any text editor and compile your code from the Linux terminal after you set up the SDK environment variables.

After you install the SDK, it copies a Bash script to the SDK directory, and running it is enough to set up environment variables for the GCC compiler and linkers.

You can source the environment variables for the current session on your host machine as follows:

```
$ source /opt/clanton-tiny/1.4.2/environment-setup-i586-poky-linux-
uclibc
```

You can check whether the setup is completed correctly for C, C++ cross compilers, and linker variables, respectively, as shown in the following code. $CC is a variable for the C compiler, $CXX is a variable for the C++ compiler, and $LD is for the Intel Galileo toolchain linker:

```
$ echo $CC

i586-poky-linux-uclibc-gcc -m32 -march=i586 --sysroot=/opt/clanton-
tiny/1.4.2/sysroots/i586-poky-linux-uclibc

$ echo $CXX

i586-poky-linux-uclibc-g++ -m32 -march=i586 --sysroot=/opt/clanton-
tiny/1.4.2/sysroots/i586-poky-linux-uclibc

$ echo $LD

i586-poky-linux-uclibc-ld --sysroot=/opt/clanton-
tiny/1.4.2/sysroots/i586-poky-linux-uclibc
```

Building applications on the Linux Terminal

Let's write a simple `Hello World` C code with a text editor, compile it on the host console, and run it on Intel Galileo using the following steps:

1. Create an empty `.c` file with your favorite text editor or any console text editor. Write a `printf` function, as follows, and compile it:

    ```
    $ nano hello_world.c
    #include <stdio.h>
    int main (void){
      printf("Hello World\n");
      return 0;
    }
    $ $CC hello_world.c -o helloworld
    ```

2. You can copy the binary file on to the Intel Galileo development board with the `scp` command. SSH and SCP tools are installed with the full image we built to boot with the SD card.

3. If the board is connected to your local network, you can easily transfer files with the `scp` command to the board from your host machine. **Secure Copy Protocol (SCP)** securely transfers computer files between a local host and a remote host. Therefore you can transfer files from Intel Galileo to your development PC or from your PC to the Intel Galileo board.

    ```
    $ scp helloworld root@192.168.2.88:/home/root/apps
    ```

     Together with the Telnet protocol and the installation of the `ssh` daemon (sshd) on Intel Galileo, you can connect to the board's terminal from the `ssh` client of your host system:

    ```
    $ ssh root@192.168.2.88
    ```

4. Open the Intel Galileo board's terminal and run the application from the folder you copied in the binary file:

    ```
    $ cd /home/root/apps
    $ ./helloworld
    Hello World
    ```

This method can be used to develop any application and deploy it to Intel Galileo. If you don't have a local network connection, it may be difficult for you to transfer files to the Intel Galileo board with a USB stick or with the SD card.

Setting up an environment to work with the Eclipse IDE

Another and more efficient way of working with embedded devices is to use the Eclipse IDE. Eclipse provides a good C/C++ development environment and remote debugging utilities for developers. Another reason to select the Eclipse IDE is that the Yocto Project has a plugin for Eclipse, and this makes it very easy to set up the SDK's location and cross-compile.

In this section, we will go through the setup process for the Eclipse IDE for development and for remote debugging.

 The Eclipse IDE requires Java to run. Make sure you have installed Java runtime on your host system.

Download the latest Eclipse IDE from its official download site at `https://www.eclipse.org/downloads/` or install it from the Ubuntu repository, as follows:

```
$ sudo apt-get install eclipse
```

If Java is installed on your host system, you can start the Eclipse IDE from the command line:

```
$ eclipse
```

Configuring the Eclipse IDE for the Yocto Project

We need to install some necessary plugins for Eclipse before installing the Yocto Project plugin on it:

1. To install new software on the Eclipse IDE, from the menu bar, go to **Help | Install New Software**.

   This will prompt you with a window to start the installation.

2. First, you need to select the download site for the plugins. On the host I use, Eclipse Kepler (the previous version is Juno) is installed so, from the **Work With** drop-down list, select the correct site.

3. Click on `http://download.eclipse.org/releases/kepler`.

   This action will load the available plugins to the following list.

   Select the required plugins to be installed. Perform the following steps to select plugins from the available software:

4. Expand **Linux Tools** and select **LTTng – Linux Tracing Toolkit**.

5. Expand **Mobile and Device Development** and select the following items from the list:

   ○ **C/C++ Remote Launch (Requires Remote System Explorer)**

   ○ **Remote System Explorer End-user Runtime**

   ○ **Remote System Explorer User Actions**

   ○ **Target Management Terminal**

   ○ **TCF Remote System Explorer add-in**

   ○ **TCF Target Explorer**

6. Expand **Programming Languages** and select **C/C++ Autotools support** and **C/C++ Development Tools**.

Then, click on the **Next >** button and read and accept the licenses agreement. This will install the plugins and restart the Eclipse IDE. The following screenshot shows you what to select from the list:

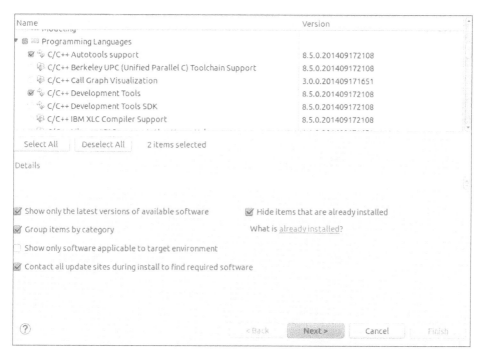

Installing the Yocto Project's Eclipse plugin

When C/C++ development and support tools for the Eclipse IDE have been installed, as described in the previous section, we are good to go with the Yocto Project's plugin installation.

You can perform the following steps to set up the Yocto Project plugin onto Eclipse:

1. Open the **Install New Software** Window again.

2. This time you should add the link for the Yocto Project plugin. Click on the **Add** button, which will open a smaller window. Enter a meaningful name (for example, `Yocto ADT Plugin`), enter `http://downloads.yoctoproject.org/releases/eclipse-plugin/1.6/kepler`, and click on **Add**.

3. When you add and select the new link from the **Work with:** menu, it will refresh the list of available packages to be installed.

4. Check the **Yocto Project ADT Plug-in**, **Yocto Project Bitbake Commander Plug-in**, and **Yocto Project Documentation plug-in** boxes.

5. Click on the **Next >** button, read and accept the license agreements, and finish the installation as shown in the following screenshot:

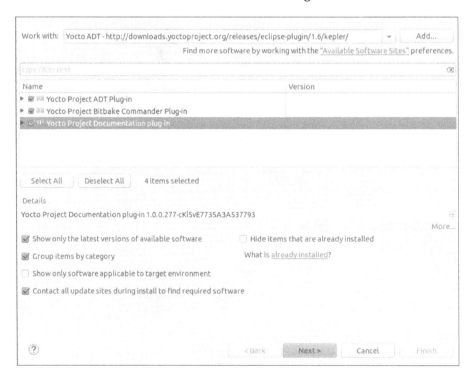

Configuring the Yocto Project's plugin for cross compilation

The Yocto plugin requires some parameters, such as the location or path of the toolchain on your host machine filesystem, to be pointed by the user.

To configure the Yocto plugin, follow these steps:

1. Click on **Window** from the Eclipse IDE menu and select **Preferences** from the list.

2. A new window will appear; from the list that appears on the left-hand side of the window, select **Yocto Project ADT**.

3. You need to enter the paths where you installed the toolchain, as shown in the following screenshot:

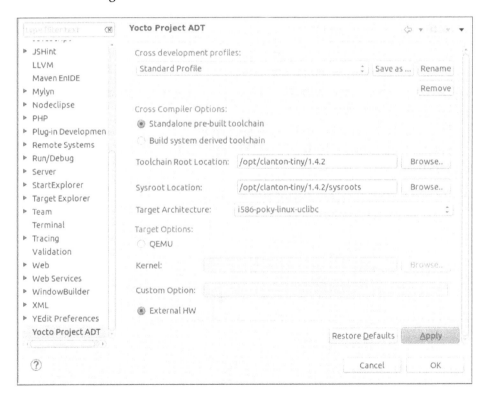

4. Click on **Apply**; if the Yocto Project plugin doesn't complain about anything, the configuration should work.

Configuring the Eclipse IDE for remote debugging

One last step in the Eclipse configuration is to configure the remote device. The Eclipse IDE makes it easy for developers to deploy and debug applications on remote devices. If we don't use a helpful tool such as Eclipse to deploy applications, we need to copy the binary to a remote device from the console or copy files to a USB stick in order to transfer the compiled binary to Intel Galileo.

The Eclipse IDE also makes it easy to debug applications on remote devices with the help of a user friendly Graphical User Interface (GUI): the **Debug** perspective. There are a couple of steps that need to be performed to complete the remote device configuration:

1. First, we will open **Remote System Explorer** in the Eclipse IDE.

2. Click on **Window** in the Eclipse IDE menu, hover over **Open Perspective** if **Remote System Explorer** is on the list. Click on it if it is not; then, click on **Other** and select **Remote System Explorer** in the opened window.

3. In the new perspective, define a new connection by clicking on the button where **Remote Systems** are listed or right-click to add a new connection.

4. This will open a new window called **Select Remote System Type**; select **Linux** and click on **Next**.

    This will show you a window, as shown in following screenshot, where you need to type in the Intel Galileo development board's IP address and a description for the connection:

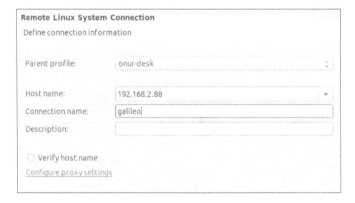

5. Select **SSH Types** in the following window when processing `.ssh` files, `processes.shell.linux`, `ssh.shells`, and `ssh.terminals`.

The default user for the remote system is inherited from the host system, so change it to root.

This process defines a remote connection. Now we need to configure Eclipse in order to run applications on a remote device, as follows:

6. In the Eclipse menu, click on **Run** and select **Run Configurations**. When the new window appears, click on the icon with a plus sign on it in order to create a new configuration.

7. You need to fill the spaces according to your host system's configuration. Select the remote connection and remote path to deploy the application.

8. If you save this instance, you can use it later for any other project you create on Eclipse to run on Intel Galileo.

The fully configured remote device is shown in the following screenshot:

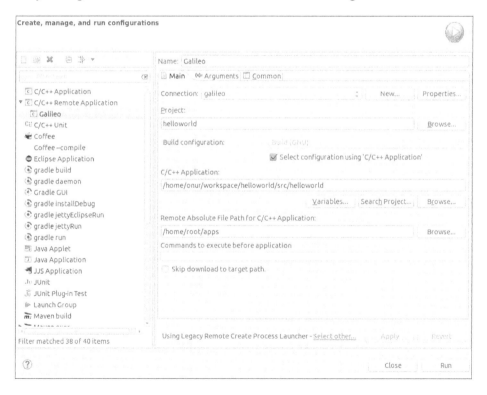

Creating a sample C project in the Eclipse IDE

We completed all the necessary configurations for Eclipse, so we can easily create an application now without reconfiguring the Eclipse IDE for the Intel Galileo toolchain. Eclipse will create C/C++ projects with toolchain configurations, and you will be able to develop C/C++ applications easily with the Eclipse IDE.

After performing a couple of easy steps, you will be able to create a C project for Intel Galileo with all the toolchain configurations, as shown here:

1. Click on **File** from the Eclipse IDE menu, select **New**, and then click on **C Project**. This will open a new window, as shown in the following screenshot.

2. Expand **Yocto Project ADT Autotools Project** and select **Hello World ANSI C Autotools Project**:

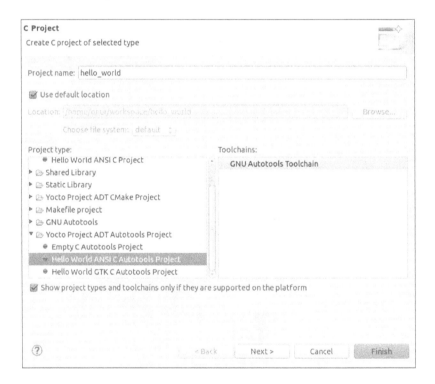

3. After you finish this, all the related files will be created. You can just build the Hello World sample by clicking on **Build** and running with the selection **Run As, Remote Application** on Intel Galileo.

Summary

In this chapter, we learned how to use the Intel Galileo development board, its software, and system development environment. It takes some time to get used to all the tools if you are not used to them. A little practice with Eclipse is very helpful to build applications and make remote connections or to write simple applications on the host console with a terminal and build them.

Let's go through all the points we have covered in this chapter. First, we read some general information about Intel Galileo and why we chose Intel Galileo, with some good reasons being Linux and the existing I/O ports on the board.

Then, we saw some more details about Intel Galileo's hardware and software specifications and understood how to work with them.

I believe understanding the internal working of Intel Galileo in building a Linux image and a kernel is a good practice, leading us to customize and run more tools on Intel Galileo.

Finally, we learned how to develop applications for Intel Galileo. First, we built an SDK and set up the development environment. There were more instructions about how to deploy the applications on Intel Galileo over a local network as well.

Then, we finished up by configuring the Eclipse IDE to quicken the development process for future development. In the next chapter, we will learn about home automation concepts and technologies.

2

Getting Started with Home Automation Applications

This chapter aims at introducing home automation concepts and existing technologies in the market along with designing our own home automation project. We will also introduce the commercially available devices for use in order to automate our home. As we learn more about home automation and existing technologies, we will also design a system with these devices and Intel Galileo.

Introducing home automation

Home automation means building a residential area, a house or a building, controlled by a centralized mechanism. Home automation can also be phrased as Smart Home, because while building a home with a centralized mechanism, we make the appliances communicate with the central hub or each other and present their current status.

In modern houses, there are many mechanical and electronic devices that need to be controlled by users. For example, if a resident needs light in their place, they will have to switch on/switch off a light bulb with the help of a switch. Another example is that you need to find out the current temperature of your house using a thermometer in order to decide whether to turn off the heater or increase the temperature. Building a system to switch it on/off remotely from a centralized control panel or switching with a condition will be the first aim of home automation.

Home automation is not limited to automating households; it is also about building a security system for your house. It is about getting alarm notifications from devices such as the door alarm if someone breaks into your home through the door, or a smoke sensor if there is a fire at your home.

Controlling appliances, such as light switches, as well as securing and surveilling your home from a controller, PC, or a similar central hub creates an automated house. We can also call it a Smart Home.

Controlling a device at home can be categorized as home automation, but the real development of home automation emerged after specific technological research and advancements. One of the first technologies was **X10**. It has been available to consumers since 1978. X10 is a protocol for communication between electronic devices. It primarily uses power line wiring for signaling and control, where the signals involve brief radio frequency bursts representing digital information.

More protocols have evolved after X10, mostly wireless protocols. **Bluetooth low energy (BLE)**, Wi-Fi, **ZigBee**, **Z-Wave**, **Insteon**, **universal powerline bus (UPB)**, **KNX**, and **EnOcean** are some of the most popular home automation protocols.

Nowadays, there are more devices that use these protocols on the market; recent technological advancements have also made these devices cheaper and allowed people to automate their homes with various of devices. Large manufacturers such as Philips, GE, Honeywell, and Schneider Electric are embracing home automation with thousands of new devices.

Together with new cool devices, the rise of cloud and mobile connection technologies (3G, LTE), and mobile devices also created user-friendly interfaces to connect and control home appliances from anywhere with an Internet connection. It is getting more and more tempting to have these cool home automation devices in your home and work with them.

An overview of home automation technologies

We just spoke about some technologies used in home automation. In this section, we will go through the details of some popular technologies used in home automation, software ecosystems, and open source projects.

Delving into home automation protocols

Numerous technologies exist for home automation, but covering all of them is not possible in this book, so we will only talk about some of the popular ones. These are the ones you will see in many retail stores or on the Internet.

There are devices for home automation that use BLE and Wi-Fi technology as well. Since they are well known and widely used technologies, we will not mention them here.

X10

X10 is a machine-to-machine communication system which was developed in 1975 by Pico Electronics Ltd. in Glenrothes, Scotland, and was the first communication protocol used for home appliances. Since wireless technologies were not as advanced as they are today, it was designed to use power line wiring in order to send signals to devices at home.

X10 uses a power line wiring system to send radio frequency signals in order to send basic digital information to devices. Digital data is encoded onto a 120 kHz carrier, which is transmitted in bursts. The X10 wireless protocol has also been developed and is used nowadays. More information about X10 can be found on its official website, http://www.x10.com.

Insteon

Insteon is a home automation technology used to define another protocol for machine-to-machine communication between home appliances. Insteon uses both wired and wireless power lines in order to carry messages to other devices in a home, just like X10 but with different encoding and frequencies. Insteon messages can carry up to 14 bytes of data at a time. More information can be found on its official website at http://www.insteon.com.

EnOcean

The EnOcean technology is mainly used in home automation but is also being used in other areas. It only transmits wireless messages to control devices.

EnOcean's wireless data packets are 14 bytes long, like Insteon, and are transmitted at 125 Kbps. The transmission frequencies used for the devices are 902 MHz, 928.35 MHz, 868.3 MHz, and 315 MHz. More information can be found on its official website at http://www.enocean.com.

Z-Wave

Z-Wave is one of the technologies we will try in order to build the examples in the following sections using Intel Galileo. The Z-Wave protocol is an interoperable wireless RF-based communication technology designed specifically for controlling, monitoring, and reading the status of devices designed for home use.

Z-Wave is a low-powered RF wireless technology that operates in the sub-1 GHz band. The lower layers, MAC and physical are described by the ITU-T G.9959 specification. It is fully backwards compatible. The Z-Wave radio uses 868.40 MHz, 869.85 MHz (Europe, South Africa, UAE, Singapore, China), 869.00 MHz (Russia), 868.10 MHz (Malaysia), 908.42 MHz (the United States), 910 MHz (Israel), 919.82 MHz (Hong Kong), 921.42 MHz (Australian/New Zealand) and 865.2 MHz (India).

There are over 1,000 interoperable products available, that is, 12 million Z-Wave products worldwide, which is one of the reasons why we have picked Z-Wave to create examples. For more information, you can visit `http://www.z-wavealliance.org` and `http://www.z-wave.com`.

ZigBee

ZigBee is also our focus. More products are being launched with ZigBee technology, such as the popular *Philips Hue* bulbs and many other devices by large manufacturers.

ZigBee is a wireless communication protocol used to create personal area networks built from small, low-power digital radios. ZigBee uses 2.4 GHz radio frequencies and is based on an IEEE 802.15 standard. It consumes little power, has a relatively nice line of sight, and transmits data over 10–100 meters depending on the power output and environmental characteristics. ZigBee has a defined rate of 250 Kbps. More information can be found at `http://www.zigbee.org`.

Introducing a home automation software ecosystem

There are numerous home automation devices, utilizing different technologies, in the market. In order to connect to a home automation device, you need a compatible controller that uses the same protocol. Some of the manufacturers only provide private APIs to connect to the hub in order to access devices.

Open source communities support some of the technologies. They develop software libraries to use USBs and serial controllers to connect and manage existing devices. You can automate your home using your personal computer with the help of open source projects. Some of the software projects that we will use in our projects are listed in the next section. You can navigate to the project's home page to get more information.

LinuxMCE

Linux Media Center Edition (LinuxMCE) is a free and open source media centered Linux distribution. LinuxMCE's main focus is to design a distribution that enables a personal video recorder, home theater PC, home automation, lighting, climate control system, surveillance and security system, and VoIP phone system with support for video conferencing systems on Linux.

LinuxMCE supports many of the home automation technologies mentioned in the previous section; you can manage the lighting and remote sockets of your home with the help of LinuxMCE.

 Check out the LinuxMCE project's home page at `http://www.linuxmce.com`.

OpenRemote

OpenRemote is a cross-platform software framework that allows you to work with many home automation protocols, as well as other commercial building automation technologies. OpenRemote supports Z-Wave, Insteon, X10, and many other protocols to help developers to automate with scripts. It is a cross-platform software that has support for Linux, Windows, Mac, and some other systems.

 You can find out more from the OpenRemote project's home page at `http://www.openremote.org`.

OpenZWave

OpenZWave is an open source software library that can be used with selected Z-Wave PC controllers to develop home automation with Z-Wave devices. It provides a useful abstraction layer for developers; you don't need to have a in-depth knowledge of the Z-Wave protocol or buy the Z-Wave development kit.

 Check out the OpenZWave project's home page at `http://www.openzwave.com`.

Other software projects

The following are some other software projects:

- **OpenHab**: This has been developed in Java and is hardware and vendor agnostic. Its aim is to provide a high-level API for developers in order to easily work with any vendor's technology on any platform. You can visit the project's home page at http://www.openhab.org.

- **Open Source Automation**: This project is open source software that runs on Windows. Developers can extend the project with a plugin to add support for any devices they want to work with. Check out the project's home page at http://www.opensourceautomation.com.

- **ago control**: This is an open source home automation solution. It provides a framework to control devices in your home. It supports many devices and protocols such as Z-Wave, X10, and some others. You can find out more from the project's home page at http://www.agocontrol.com.

- **HomeGenie**: This is an open source home automation solution. It has been designed as a home server. It can be customized according to your needs. HomeGenie can interface with X10, Insteon, Z-Wave, Philips Hue, UPnP, and RFXCom devices, as well as communicate with external web services and integrate all of this into a common automation environment. Check out the project's home page at http://www.homegenie.it.

Home automation devices, sensors, and controllers

Before designing our home automation project, let's take a closer look at some existing devices that we use for the projects in the following chapters.

There are many devices enabled with the technologies already mentioned. In order to use these devices, they have to be remotely controllable, and some data needs to be sent to the central controller with some kind of protocol. If a wall plug is not enabled with any of the home automation technologies, you can't reach the device to switch it on or off. If a wall plug has an energy meter to measure how much energy is consumed, the controller should be able to poll data.

If we are building our own hobby project to build a device for home automation, this new device should be controllable and should be able to send and receive data with a particular protocol.

Some sample devices with their descriptions and use for home automation are listed in the following table. Some are used for energy management, while some are used for security.

A Z-Wave USB controller A home automation system requires a controller device, a PC, or Intel Galileo to send/receive information from the devices. This is one of the sample controllers, easily used with any device that has a USB host on board. Other technologies such as Insteon, X10, and so on also have similar devices. Host devices use serial communications through the USB connection, and with the help of this controller, you can connect compatible devices from your PC or similar board. This device, Aeon Labs Z-Stick, is also supported by the open source projects already mentioned.	
Wall plug The image shows a plug manufactured by a company called Fibaro. It uses the wireless Z-Wave protocol to send and receive commands. Energy usage can also be monitored and fetched from the plug. Similar devices are also available with other protocols such as, ZigBee, X10, and so on. Using a switch such as this helps you manage any home appliance plugged into this plug.	

Motion and door sensor A company called Philio produces the device shown opposite. The device includes a door/window sensor and a motion sensor. It is also a Z-Wave device used to get information from sensors. This kind of device can be used for multiple purposes. It can be used for energy management, such as implementing an application to switch on lights when motion is detected. Another use case is when a window or door is open for too long, it can switch off the heater to save energy. Besides energy management, they are also useful for security. If someone breaks into your home, you can detect it. Similar devices are also available for ZigBee, Insteon, and other protocols.	
Water leak sensor Water leak sensors can be categorized into the security and surveillance part. If there is a flood, or water in a place where there isn't supposed to be any, the sensor alerts the user or the home automation system and all electricity switches can be closed. Everspring produces the device shown opposite. Similar devices are available from other manufacturers and with other protocols as well. Smoke and carbon monoxide (CO) detectors also work on similar principles. If there is a change in the level of smoke or CO, an alarm will be raised and the resident can take the necessary precautions.	

Designing a home automation project with Intel Galileo

It's time for us to combine our knowledge of Intel Galileo and of the previous sections of this chapter to design a Smart Home project.

We know that a Linux operating system is running on Intel Galileo and Intel Galileo has a USB host port, together with Arduino-compatible pins. We can use the USB controller on Intel Galileo and connect to home devices remotely. After establishing a communication link with these devices, we can send and receive data.

We will use a USB Z-Wave controller to connect to Z-Wave devices wirelessly. This will teach us how to communicate with Smart Home appliances and convert Intel Galileo into a Smart Home hub.

The Intel Galileo pinout support includes digital and analog input pins as well as I2C, SPI protocols. We can use widely available temperature sensors such as LM35 and LM74 with Intel Galileo. You can find some temperature sensors from Texas Instruments from `http://www.ti.com/lsds/ti/analog/sensors/overview.page`.

Temperature sensors will help you gather temperature data from the environment. We can use these values to control other appliances that we have already connected. For example, if the temperature is high and we have connected our wall plug to the heater, our application can switch off the wall plug and the heater will stop working.

Another utility of Intel Galileo is that it has an Ethernet plug and we can connect mini PCI-E Wi-Fi cards to add a wireless connection capability to it. Connecting Intel Galileo to the local network will enable you to monitor your house using other devices that are connected to the same network. Node.js and Python can be run on Intel Galileo. Adding Node.js and Python to Intel Galileo will require more space, so we can use an SD card image, which we described in the *Chapter 1, Getting Started with Intel Galileo*. Then, we can implement and run a simple web server to control and host data on Intel Galileo and create a user-friendly web interface to access it.

The following figure shows our design of a smart home with Intel Galileo. The Z-Wave controller, remote Z-Wave devices, temperature and humidity sensors, and IP camera are the devices we will use for our project. The entire system is connected to the home network and can be monitored from any device that is connected to the local network.

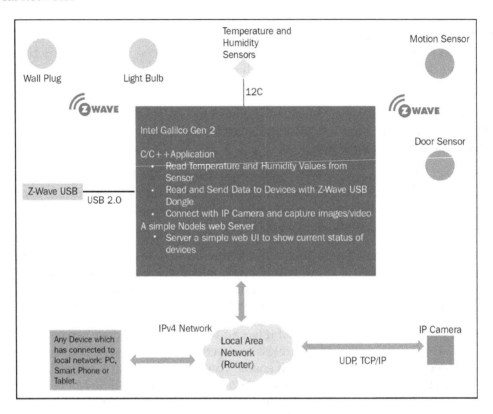

In the preceding figure, we can see the basic setup of our home automation system. Implementing it may take some time. In the following chapters, you will be shown the critical information and development steps necessary to complete our project.

We will start by using temperature and humidity sensors with Intel Galileo. We will implement a C project to read data from the sensors. This will be a very informative section to understand Intel Galileo's connection to sensors via the Arduino pins.

Then, we will set up a Z-Wave USB to Intel Galileo and create a serial communication within the application. When we have successfully set up the Z-Wave USB controller, we will try to reach out to the remote devices and send and receive data with the Z-Wave protocol.

We will also have a Wi-Fi-and Ethernet connection-enabled RGB camera, which will be able to stream images and video to the network. We will also try to set up a connection with a camera in our application, and we will learn how to connect as well as capture images and videos from the device.

> Network-enabled (Wi-Fi) IP cameras are widely used for security. They are also called CCTV cameras. A common use case is for parents who monitor their infants when they are sleeping.

There are hundreds of devices in the market, some of which are available for use with our controller and some of which are not enabled for use with our controller. With the correct selection of devices and controller, you can build any home automation project with the chosen technology. In this book, we will only be able to cover the commonly used devices, in order to understand the concepts behind home automation and gain the knowledge to extend our projects.

Now, it is time to create a new C application project in the Eclipse IDE, as we did in the previous chapter.

Summary

In this chapter, we covered home automation and some details of existing technologies.

First, we defined the home automation concept. Then, we followed up by looking at existing technologies used to build Smart Homes. We introduced X10, Insteon, EnOcean, Z-Wave, and ZigBee, along with links to their official web pages.

We also spoke about open source software projects for home automation. We went through some of them and found out useful information about these projects. Then, we proceeded to discuss a couple of existing devices used for home automation.

Finally, using this knowledge of the relevant technologies and devices, we designed our home automation project.

Now, it is time to play with some home automation gadgets in order to learn more about Intel Galileo. In the following chapter, we will make some small projects with Intel Galileo to measure the temperature and humidity of our home with a sensor connected to Intel Galileo. Then, we will communicate with a wall plug using the Z-Wave protocol to improve the energy management of our home.

3

Energy Management with Environmental and Electrical Sensors

Energy management is one of the major purposes of home automation together with home security and surveillance. Energy management means saving and using energy efficiently in residential areas so that you can reduce electricity consumption and create a green, energy saving house.

In this chapter, we will learn how to connect sensors to Intel Galileo and develop C applications that will give us the ability to develop specific devices only with Intel Galileo.

After you learn how to develop applications for Intel Galileo with a sample application, we will follow up with controlling electrical switches remotely to control the energy consumption of our home.

Delving into energy management

Energy management is about managing devices that consume energy in order to save energy and control their consumption. Saving energy will lead to the more efficient usage of resources and will reduce consumption. For energy management in Smart Homes, there are sensors that help you control devices to power them on or off.

Many sensors and devices are being produced for energy management purposes. Temperature sensors and energy meters are key sensors; relay switches and dimmers are essential actuators of energy management systems. You can either get a sensor and develop your own device to embed into your home, or you can directly buy devices to embed into your home and control them.

There are wall plugs with both energy meters and on/off switches, and lamp holders with switches and dimmers. Other sensors can be integrated with the energy management system to make them work more efficiently and automatically. Temperature, motion, or light sensors can be used to decide whether to switch on or switch off a remote switch on the system.

In the chapter, we will start with a simple sensor to see how we can connect it to Intel Galileo and get readings from it. With what you have learned from this section, you can proceed to work with more complex sensors. Then we will proceed with the Z-Wave controller, which is connected to a remote wall plug to switch it off.

Developing sensor-based applications with Intel Galileo

Let's start with learning how to use sensors with Intel Galileo. In this section, we will develop an application with Intel Galileo. We pick the temperature and humidity sensors for application development because knowing the temperature of your house is the key to energy management. With temperature data, you can decide whether to switch on or switch off the heater or air conditioner in your home to save energy.

In order to get temperature data, we will use a new device: the *Sensirion SHT11 Temperature and Humidity Sensor*. We will connect this sensor to Intel Galileo's pinout to read temperature and relative humidity values. All the development will be done with the C programming language and Linux operating system, as mentioned earlier.

When we are finished with this section, you will have the knowledge to use pins on Intel Galileo and GPIO device files in the Linux filesystem. These methods can be reused for other sensors and devices in your future applications with them.

Understanding the working of a sensor from the datasheet

We have picked a sensor, that uses a data and clock line to enable measurement on sensors. There are many temperature sensors in the market that are compatible with Intel Galileo. LM35 is a widely used temperature sensor, and it provides a voltage output proportional to Centigrade temperatures. Our SHT11 sensor provides us with digital data as raw temperatures and provides the coefficients to convert the raw data to human- readable Centigrade temperatures.

The image shown here is a SHT11 sensor:

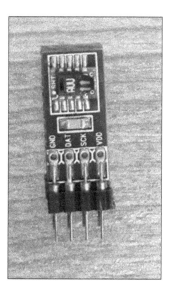

While working with sensors on Intel Galileo or any other platform, we need to first read the datasheet provided by the manufacturer to understand the working of the sensor and its operational limits.

 The datasheet for the SHT11 is accessible from the following link: http://www.sensirion.com/fileadmin/user_upload/ customers/sensirion/Dokumente/Humidity/Sensirion_ Humidity_SHT1x_Datasheet_V5.pdf

Understanding the critical information provided on datasheets is essential. As we have read in the sensor datasheet, SHT11 works with simple clock and data lines to read and write bits. It doesn't use any specific bus or protocol such as I2C, SPI, or UART, and so we will pick two GPIO pins from Intel Galileo to send and get bits to/from the SHT11 sensor.

To start reading from the sensor, we need to send a byte as a command to SHT11. Our application will send 0x03 bytes to SHT11 in order to measure temperature and 0x05 bytes to measure relative humidity; you can find the commands in the datasheet. Since the sensor is not using any communication bus, we will use GPIO pins to send and get data bit by bit. The figure given here shows how we send bits from a DAT pin.

The same method can be used to read from a DAT pin.

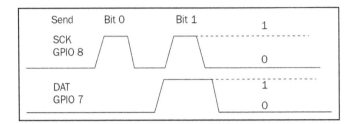

Before sending the command, there are a few steps that need to be carried out with clock data to tell the sensor about the incoming command. After sending the command, it is also required to complete some bit transfers and readings from the sensor to acknowledge that the command was received. The following figure shows the steps to read temperature data from the SHT11 sensor:

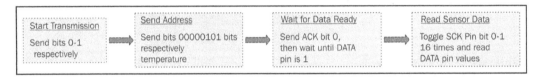

The last step is to wait for the sensor to make data pin low (0) to start reading bits from that. In order to complete the temperature measurement from the sensor, we need to complete all the steps described. For more details, you can follow up with the sensor datasheet.

Connecting our sensor to Intel Galileo

Let's connect our sensor to Intel Galileo before coding. The **SHT11** sensor has four pins to be connected: the VCC, GND, clock (SCK), and data (DAT) pins.

We will use 5V for **VCC** since it is operational with 5V and produces more accurate data. The **GND Pin** can easily be seen on the Intel Galileo pinout. For the clock and data, we picked two GPIO pins **IO7** and **IO8** to send 0 and 1 to the sensor. **IO7** on the board will be the data line and **IO8** will be the clock line of the **SHT11** sensor.

The following figure basically represents the pin connections of **SHT11** to the corresponding pinout on Intel Galileo.

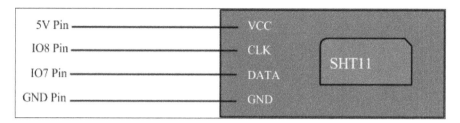

Starting application development with C

We have connected SHT11 to Intel Galileo. Now we can start the development of our application to read environmental temperature and humidity values. As described in *Chapter 1, Getting Started with Intel Galileo*, let's create a new C project and name it as thermostat.c or thermometer.c. It will create a C file named after your project with a main function.

First, we need to include the following headers and define the constants as needed:

```
#include <stdlib.h>
#include <stdio.h>
#include <fcntl.h>
#include <stdint.h>
#define LSBFIRST 0
#define MSBFIRST 1
// Global Definitions for High/Low and Input/Output values for
//GPIO Operations
#define HIGH 1
#define LOW 0
#define INPUT "in"
#define OUTPUT "out"
//Buffers needed for I/O Operations
#define BUF 8
#define MAX_BUF 256
// IO7 and IO8 pins Linux GPIO numbers to access from Linux
#define DATA_PIN 38  //IO 7
#define CLOCK_PIN 40 //IO 8
//SHT11 Commands Defined in Datasheed
```

```
const uint8_t measure_temperature = 0b00000011;
const uint8_t measure_humidity = 0b00000101;
const uint8_t read_status = 0b00000111;
const uint8_t write_status = 0b00000110;


//Temperature Measurement Constants as in Datasheet
// T =   d1 + d2 * SOt
const float D1_5V = -40.1;
const float D2_14bit = 0.01;
const float D2_12bit = 0.04;
//Relative Humidity Measurement Constants as in Datasheet
// RH = c1 + c2 * SOrh + c3 * SOrh * SOrh
const float C1 = -2.0468;
const float C2 = 0.0367;
const float C3 = -0.00395484;


//Global Variables to Store Temperature and Relative Humidity
float temperature = 0.0; //Celcius
float humidity = 0.0; //Relative Humidity
```

We will send bits by toggling GPIO pins and read bits from GPIO by setting the GPIO direction as the input. GPIO pins direction in the way it is working; if GPIO is defined as the output, it can send 0 or 1 to pinout; if it is defined as input, a low, or high signal coming to the pin can be read. In the Linux operating system, everything is a file, virtual or real, and so you can access virtual files to read and write data. Each pin on Intel Galileo has a corresponding device file on the Linux filesystem. Here you can find all corresponding numbers from the Intel Galileo I/O Mapping document. **IO7** and **IO8** are GPIO 38 and 40 respectively on the Linux filesystem.

 The Intel Galileo I/O Mapping document is available at `http://www.intel.com/support/galileo/sb/CS-035205.htm`

Let's continue with our `main` function. We will simply read the temperature at first like this:

```
int main(void) {
  read_temperature();
  printf("Temperature is %f Centigrade\n", temperature);
  return 0;
}
void read_temperature(){
// Conversion coefficients from SHT11 datasheet
  const float D1 = -40.0;    // for 14 Bit @ 5V
  const float D2 = 0.01; // for 14 Bit DEGF
```

```
    send_command_SHT11(measure_temperature);
    wait_for_result();
    int raw = retreive_data_SHT11();
    skip_crc();
    temperature = 0.0;
    temperature = ((float) raw * D2) - D1;
}
```

The `read_temperature` function first sends a command to the SHT11 sensor bit by bit, waits for the sensor to take the data line to low and then retrieves the raw temperature value from the sensor. The first data read from the sensor is the raw value (in the datasheet, Celsius and Fahrenheit degree calculation equations and coefficients are given); finally we apply the equation and get the result.

Let's proceed to the `send_command_SHT11` function:

```
int send_command_SHT11(int command) {
  int ack;
  //Start Transmission
  gpio_set_mode(DATA_PIN, OUTPUT);
  gpio_set_mode(CLOCK_PIN, OUTPUT);
  gpio_set_value(DATA_PIN, HIGH);
  gpio_set_value(CLOCK_PIN, HIGH);
  gpio_set_value(DATA_PIN, LOW);
  gpio_set_value(CLOCK_PIN, LOW);
  gpio_set_value(CLOCK_PIN, HIGH);
  gpio_set_value(DATA_PIN, HIGH);
  gpio_set_value(CLOCK_PIN, LOW);
  //Shift Out
  shiftOut(MSBFIRST, command);
  gpio_set_value(CLOCK_PIN, HIGH);
  gpio_set_mode(DATA_PIN, INPUT);
  delay(20);
  ack = gpio_get_value(DATA_PIN);
  if (ack != LOW) {
    perror("ACK Error 0");
  }
  gpio_set_value(CLOCK_PIN, LOW);
  delay(320);
  ack = gpio_get_value(DATA_PIN);
  if (ack != HIGH) {
    perror("ACK Error 1");
    gpio_set_mode(DATA_PIN, OUTPUT);
    gpio_set_value(DATA_PIN, HIGH);
    gpio_set_mode(DATA_PIN, INPUT);
  }
  return 0;
}
```

The `send_command_SHT11` function has been implemented according to the datasheet. In the function code, before sending the given command, we need to first make the sensor ready by sending 0s and 1s from CLK and DATA GPIO pins. The order of this data is given in the datasheet. After we make the initialization, we will send the command starting from the leftmost bit (MSB). For the temperature measurement command (0x00000011), we will send first six 0s then two 1s. The Shift Out command toggles the GPIO pins according to the given direction (MSBFIRST or LSBFIRST).

Let's now proceed with the way we handled GPIO pins on the Intel Galileo. We have the following four functions to handle GPIO operations:

```
int gpio_export(int gpio_num);

int gpio_set_mode(int gpio_num, const char* mode);

int gpio_set_value(int gpio_num, int value);

int gpio_get_value(int gpio_num);
```

The GPIO device files are in directory `/sys/class/gpio`. Most of the GPIO device files do not exist by default; in order to create the device files, we will use the `gpio_export` function. When the device exists, we will set the mode of the GPIO according to our need for input or output with, `gpio_set_mode` function. Then, with `gpio_set_value` and `gpio_get_value`, we will set and read the value of GPIO pins as shown here:

```
int gpio_export(int gpio_num) {
  //Device File Path Declarations
  const char* gpio_export = "/sys/class/gpio/export";
  //Device File Declarations
  int fd_x = 0, g_err = -1;
  //Buffer
  char g_buf[BUF];
  fd_x = open(gpio_export, O_WRONLY);
  if (fd_x < 0) {
    printf("Couldn't get export FD\n");
    return g_err;
  }
  //Export GPIO Pin
  sprintf(g_buf, "%d", gpio_num);
  if (write(fd_x, g_buf, sizeof(g_buf)) == g_err) {
    printf("Couldn't export GPIO %d\n", gpio_num);
    close(fd_x);
    return g_err;
  }
```

```
    close(fd_x);
    return 0;
}
```

The export function opens the /sys/class/gpio/export device file to create
requested GPIO pin device files as shown here:

```
int gpio_set_mode(int gpio_num, const char* mode) {
  //Device Direction File Path Declarations
  const char* gpio_direction_path =
"/sys/class/gpio/gpio%d/direction";
  //Device File Declarations
  int fd_d = 0, g_err = -1;
  //Buffers
  char pindirection_buf[MAX_BUF];
  char d_buf[BUF];
  //Set pin number and set gpio path
  if (sprintf(pindirection_buf, gpio_direction_path, gpio_num) <
0) {
    printf("Can't create pin direction file path\n");
    return g_err;
  }
  //Open GPIO Direction File
  fd_d = open(pindirection_buf, O_WRONLY);
  //If GPIO doesn't exist then export gpio pins
  if (fd_d < 0) {
    if (gpio_export(gpio_num) < 0) {
      return g_err;
    }
    fd_d = open(pindirection_buf, O_WRONLY);
    if (fd_d <= 0) {
      printf("Couldn't get direction File for pin %d\n",
gpio_num);
      return g_err;
    }
  }
  sprintf(d_buf, mode);
  if (write(fd_d, d_buf, sizeof(d_buf)) == g_err) {
    printf("Couldn't set direction for pin %d\n", gpio_num);
    return g_err;
  }
  close(fd_d);
  return 0;
}
```

The gpio_set_mode function sets the GPIO pin as input or output as requested. For example, in order to set GPIO pin 38 as output, the function opens the /sys/class/gpio/gpio38/direction file and writes out to file and then GPIO 38 will start working to provide the following output:

```
int gpio_set_value(int gpio_num, int value) {
  //Device Direction File Path Declarations
  const char* gpio_value_path = "/sys/class/gpio/gpio%d/value";
  //Device File Declarations
  int fd_v = 0, g_err = -1;
  //Buffers
  char pinvalue_buf[MAX_BUF];
  char v_buf[BUF];
  //Set pin number and set gpio path
  if (sprintf(pinvalue_buf, gpio_value_path, gpio_num) < 0) {
    printf("Can't create pin direction file path\n");
    return g_err;
  }
  //Open GPIO Value File
  fd_v = open(pinvalue_buf, O_WRONLY);
  //If GPIO doesn't exist then export gpio pins
  if (fd_v < 0) {
    if (gpio_export(gpio_num) < 0) {
      return g_err;
    }
    fd_v = open(pinvalue_buf, O_WRONLY);
    if (fd_v <= 0) {
      printf("Couldn't get value File for pin %d\n", gpio_num);
      return g_err;
    }
  }
  sprintf(v_buf, "%d", value);
  if (write(fd_v, v_buf, sizeof(v_buf)) == g_err) {
    printf("Couldn't set value for pin %d\n", gpio_num);
    return g_err;
  }
  close(fd_v);
  return 0;
}
```

The gpio_set_value function sets the value of GPIO as shown in the preceding code. The function gets the GPIO value file and writes the given value to the device file. As an example, for GPIO 38, the value file is stored at /sys/class/gpio/ gpio38/value, as shown here:

```
int gpio_get_value(int gpio_num) {
  //Device Direction File Path Declarations
  const char* gpio_value_path = "/sys/class/gpio/gpio%d/value";
  //Device File Declarations
  int fd_v = 0, g_err = -1;
  //Buffers
  char pinvalue_buf[MAX_BUF];
  char v_buf[BUF];
  //Set pin number and set gpio path
  if (sprintf(pinvalue_buf, gpio_value_path, gpio_num) < 0) {
    printf("Can't create pin direction file path\n");
    return g_err;
  }
  //Open GPIO Value File
  fd_v = open(pinvalue_buf, O_RDONLY);
  //If GPIO doesn't exist then export gpio pins
  if (fd_v < 0) {
    if (gpio_export(gpio_num) < 0) {
      return g_err;
    }
    fd_v = open(pinvalue_buf, O_RDONLY);
    if (fd_v <= 0) {
      printf("Couldn't get value File for pin %d\n", gpio_num);
      return g_err;
    }
  }
  if (read(fd_v, v_buf, 1) == g_err) {
    printf("Couldn't get value for pin %d\n", gpio_num);
    return g_err;
  }
  close(fd_v);
  return atoi(v_buf);
}
```

The gpio_get_value function is used to get the value of input mode GPIO. When a GPIO is used as input and a high signal is given to the corresponding GPIO, it will return as 1. This function simply gets the GPIO number and returns the value for the user.

GPIO functions simply teach you how to work with device files in the Linux operating system. When the protocol gets more complex, you will need to make more configurations, but you will also be able to use open source Linux libraries to read and write devices.

We also have two other functions to send or read bytes from GPIOs. The shiftOut function helps us to write multiple bits at one time with the given directions and the shiftIn function will help to read multiple bits as a byte, as shown here:

```c
void shiftOut(uint8_t bitOrder, uint8_t val) {
  uint8_t i;
  for (i = 0; i < 8; i++) {
    if (bitOrder == LSBFIRST) {
      gpio_set_value(DATA_PIN, !!(val & (1 << i)));
    } else {
      gpio_set_value(DATA_PIN, !!(val & (1 << ((8 - 1 - i)))));
      gpio_set_value(CLOCK_PIN, HIGH);
      delayMicroseconds(80);
      gpio_set_value(CLOCK_PIN, LOW);
    }
  }
}
int shiftIn(int bit_order, int n_bits) {
  int ret = 0;
  int i;
  gpio_set_value(CLOCK_PIN, LOW);
  for (i = 0; i < n_bits; ++i) {
    gpio_set_value(CLOCK_PIN, HIGH);
    if (bit_order == LSBFIRST) {
      ret |= gpio_get_value(DATA_PIN) << i;
    } else {
      ret |= gpio_get_value(DATA_PIN) << n_bits - i;
    }
    delayMicroseconds(20);
    gpio_set_value(CLOCK_PIN, LOW);
  }
  return (ret);
}
```

You may notice that there are functions to delay clock bit setting to get the sensor ready. These functions are implemented with `usleep` function from the standard C library.

As we have seen in the `read_temperature` function, first we send our command and then wait until the data is ready for checking. If the data bit is set to 0, the `wait_for_result` function does the checking. Then we retrieve the data as shown in the following `retrieve_data_SHT11` function:

```
void wait_for_result() {
  int ack;
  gpio_set_mode(DATA_PIN, INPUT);
  int i;
  for (i = 0; i < 100; i++) {
    delay(20);
    ack = gpio_get_value(DATA_PIN);
    if (ack == LOW) {
      printf("Sensor Data Ready\n");
      break;
    }
  }
  if (ack == HIGH) {
    perror("ACK error 2");
  }
}
int retreive_data_SHT11() {
  int16_t value;
  gpio_set_mode(DATA_PIN, INPUT);
  gpio_set_mode(CLOCK_PIN, OUTPUT);
//Read first 8 bits
  value = shiftIn(MSBFIRST, 8);
  //Shift 8 bit to right
  value *= 256;
  //Send Acknowledgment
  gpio_set_mode(DATA_PIN, OUTPUT);
  gpio_set_value(DATA_PIN, HIGH);
  gpio_set_value(DATA_PIN, LOW);
  gpio_set_value(CLOCK_PIN, HIGH);
  gpio_set_value(CLOCK_PIN, LOW);
  //Get MSB
  gpio_set_mode(DATA_PIN, INPUT);
  value |= shiftIn(MSBFIRST, 8);
  return value;
}
```

To read data from the sensor, we need to read the first 8 bits, send an acknowledgment bit, and then read the last 8 bits to get 16-bit raw data. Since we are reading bits in left-to-right order, we need to shift the first 8 bits to the left by 8. You can do this either by multiplying by 256 or using the bitwise operator <<. After reading the first 8 bits, we will read 1 bit and shift them to the left when the new bit has been read. And finally, we will have 16-bit raw data to get the temperature or relative humidity.

> The skip_crc function sends three bits to skip the checksum. We just want to get raw data quickly; if you install a real system, you may want to use the sensor's checksum utility. Please refer to the datasheet for this.

We have defined all the functions in the C file we created in the Eclipse project. It is time to build and deploy it to Intel Galileo. After we deploy and run the application from the command line or through Eclipse IDE, it will prompt the following output with respect to the measured temperature:

```
root@clanton:~# ./thermometer

Temperature is 20.3 Centigrade
```

So, we have completed our first sample application, and we get the temperature of the environment where the sensor is placed. By using the functions defined in this section, we can implement the function to measure relative humidity as well.

We need to send the humidity measure command instead of the temperature command using send_command_SHT11. Then, we need to calculate relative humidity with the provided coefficient and equation from the sensor datasheet. The following function can be used to read humidity values from the SHT11 sensor:

```
void read_humidity() {
    send_command_SHT11(measure_temperature);
    wait_for_result();
    int hum = retreive_data_SHT11();
    skip_crc();
    humidity = 0.0;
    humidity = -4.0 + (0.0405 * (float) hum)
        + (-0.0000028 * (float) hum * (float) hum);
}
```

Within this section, we have worked with a sensor to produce our own thermometer. Having temperature data is critical for decision-making while saving energy. Let's say you have a remote plug connected to your heater; if the temperature value has reached the degree you want, you can close the heater by turning off the wall plug remotely. In the following section, we will look into remote switches and try to control one remotely with Z-Wave controller.

Energy management with remote switches

Remote switches are another key component required to automate your home. A remote switch gives the ability to remotely switch on and switch off if the plugged device is required to be open or closed. Some remote switches also have energy meters on them that provide the amount of consumed energy to the user.

We have a remote wall plug from Fibaro that uses the Z-Wave protocol to communicate. The Fibaro wall plug has a relay switch to turn it on/off and an energy meter to provide a power consumption value. You can see the device in the following image:

We will try to switch it on and off by sending basic commands through a Z-Wave controller. We will use the **Aeon Labs Z-Stick S2** USB adapter for our controller. As Intel Galileo has a USB host port, we will connect the Z-Wave controller from Intel Galileo's USB host. The Z-Wave USB adapter uses serial communication when it's on a host device, and so we will implement a couple of C functions to create communication with the device. You can see the USB adapter in the following image:

While we are trying to work with Z-Wave USB adapter, we will learn some new features to work with Intel Galileo.

Z-Wave adapters need to include the device before you program it. The inclusion steps for devices are covered in the device manuals. Either you can read them from the paper provided in the product box or you can access them from their official web sites.

The Aeon Labs Z-Stick S2 manual is available from here: `http://aeotec.com/z-wave-usb-stick/913-z-stick-manual-instructions.html`

The Fibaro wall plug manual is available from here: `http://www.fibaro.com/manuals/en/FGWPE_F-101-Wall-Plug/FGWPE_F-101-Wall-Plug-en.pdf`

The Aeon Labs Z-Stick requires the cp210x kernel module to be loaded to create serial communication device: `/dev/ttyUSB0`.

Building kernel modules for Intel Galileo

We require the **cp210x** kernel module for our Intel Galileo image. In order to do that, we will go to our build system with Yocto Project and build the cp210x kernel module; we'll then install it into Intel Galileo using the following steps:

1. Go to the directory that you created for building the SD card image for Intel Galileo. This is shown here:

   ```
   $ cd /home/onur/galileo_build/meta-clanton_v1.0.1
   ```

2. We need to set up the environment variables again. If your terminal session is still open from the first chapter, you don't need to rerun this command. This is shown here:

   ```
   $ source poky/oe-init-build-env yocto_build
   ```

3. Let's reconfigure kernel and pick the cp210x module to build. This is shown here:

   ```
   $ bitbake linux-yocto-clanton -c menuconfig
   ```

   The last command will open the kernel configuration menu window for you. In order to select cp210x module, select **Device Drivers**, **USB Support**, **USB Serial Converter Support** and finally **USB CP210x** Family of UART from the menu; save the configuration and exit.

4. Build the kernel module with the following command:

   ```
   $ bitbake linux-yocto-clanton
   ```

   The final command will build the module and create the installable package in `yocto_build/tmp/deploy/ipk/clanton` directory named `kernel-module-cp210x_3.8-r0_clanton.ipk`.

5. Copy the file to Intel Galileo; install and load the kernel module as shown here:

   ```
   $ scp kernel-module-cp210x_3.8-r0_clanton.ipk
   root@192.168.2.88:/home/root
   ```

   ```
   $ opkg install kernel-module-cp210x_3.8-r0_clanton.ipk
   ```

6. The final command will install the module to the `/lib/modules/kernel/drivers/usb/serial` directory. Load module to Intel Galileo with this command:

   ```
   $ modprobe /lib/modules/kernel/drivers/usb/serial/cp210x
   ```

When you reboot, it won't load the module; if you want to automatically load the module, edit the `/etc/modules-load.quark/galileo_gen2.conf` file and add cp210x to the file.

Serial communication on Linux

Serial communication is a widely used protocol to receive and transfer bytes from/to devices. For example, the 6-pin FTDI cable you've connected to your Intel Galileo and your PC's USB host communicate using the RS232 serial protocol, to send the keyboard data you entered on minicom and get the terminal output to your device.

Most of the PC interfaces of home automation controllers use serial communication. We will use the Aeon Labs Z-Stick, but there are other USB adapters that can be found for use with Z-Wave devices.

The Insteon and EnOcean USB adapters are also available, and can use serial communication as well. Refer to the Insteon USB Interface here: http://www.insteon.com/2413U-PowerLinc-USB.html.

First we need to set up the serial communication interface from our application. We will continue with a new project only with this purpose in mind. First, we will open the device file `/dev/ttyUSB0` and then we will set the serial interface with a 1,15,200 baud rate, 8 data bits, 1 stop bit, and no parity bits. If this setting isn't done right, the sent or received data will be corrupted.

We defined the function `open_serial_device` to get the device file and set the device communication options as shown here:

```
int open_serial_device(const char* serial_device_path) {
 //Open Device File /dev/ttyUSB0
 int device_file = open(serial_device_path, O_RDWR | O_NOCTTY, 0);
 if (device_file < 0) {
    printf("Can't Open Serial Controller\n");
    return -1;
  }
struct termios options;
 int bits;
  bits = 0;
  ioctl(device_file, TIOCMSET, &bits);
  /**
```

```
 * Get Current Options
 */
tcgetattr(device_file, &options);
//No Parity
options.c_iflag = IGNPAR;
options.c_iflag |= IGNBRK;
//Set 8 Data bits
options.c_cflag |= CS8 | CREAD | CLOCAL;
options.c_oflag = 0;
options.c_lflag = 0;
int i;
for (i = 0; i < NCCS; i++) {
  options.c_cc[i] = 0;
}
options.c_cc[VMIN] = 0;
options.c_cc[VTIME] = 1;
//Set Baudrate for Serial Communication
cfsetspeed(&options, B115200);
if (tcsetattr(device_file, TCSANOW, &options) < 0) {
  printf("Can't Set The Serial Device Parameters\n");
} else {
  printf("Successfully Set the Serial Device Parameters\n");
}
//Flush the waiting bits/bytes
tcflush(device_file, TCIOFLUSH);
return device_file;
}
```

In the function, we first open the serial device file, then we request the current options from system function `tcgetattr` and complete the settings to make the device ready for communication.

Now, we will proceed to implement the methods to read and write bytes from/to the serial device using the `read_from_serial_device` and `write_to_serial_device` functions.

The following function writes bytes to the serial device and then checks whether the bytes are written correctly:

```
int write_to_serial_device(int device_file, uint8_t buffer[], int
length) {
  int i = 0;
  int count = 0;
  printf("Writing: ");
  while (count < length) {
    if (write(device_file, &buffer[count], 1) == 1) {
```

```
        printf("0x%x ", buffer[count]);
        count++;
    } else {
        printf("\n Can't Write Byte\n");
        break;
    }
  }
  if (count == length) {
    printf("\nWrite Successful\n");
    return count;
  } else {
    printf("\nCan't Write All Frame to Serial Device\n");
    return -1;
  }
}
```

The following function read bytes from the serial device:

```
int read_from_serial_device(int device_file, uint8_t *data) {
  int bytesRead;
  uint8_t buffer[256];
  //Read Data From Serial Device to Buffer
  bytesRead = read(device_file, buffer, sizeof(buffer));
  //Print Bytes
  data = malloc(sizeof(uint8_t) * bytesRead);
  if (bytesRead > 0) {
    int i = 0;
    printf("Received: ");
    for (; i < bytesRead; i++) {
      data[i] = buffer[i];
      printf("0x%x, ", data[i]);
    }
    printf("\n");
  }
  return bytesRead;
}
```

Controlling a remote wall plug from Intel Galileo

So we have implemented the required functions for serial communication. In order to communicate correctly with the Z-Wave controller, we need to learn basic Z-Wave commands to get and set values from/to the remote wall plug. We will obtain help from open source projects to get information about Z-Wave communication.

In the following sample, we will only use basic commands to switch on/off and read energy meter value from the Fibaro wall plug. We will not parse the raw data received from the Z-Wave controller. In the following chapter, we will parse the received values and create a more automated application by bringing together the blocks.

Basically, the Z-Wave controller creates a home network and generates a number (or a home ID) for that. In this way, it isolates your devices from any other controller to reach out and give a number for each of the devices or nodes added, starting from 2. The first node is the controller itself.

The following function is the `main` function that is used to simply get power information from the wall plug. We create a thread to read the received data while we send data to the controller.

```c
void* reader_thread(void *arg) {
  while (1) {
    read_from_serial_device();
  }
  return NULL;
}
int main(int argc, char* argv[]){
//Open Aeon Stick Z-Wave
 device_file = open_serial_device(SERIAL_DEVICE);
 if (device_file < 0) {
  return EXIT_FAILURE;
 }
  //Create Thread
  pthread_t reader;
  int err = pthread_create(&reader, NULL, &reader_thread, NULL);
  if (err != 0) {
    close_serial_device();
    printf("Can't create Thread\n");
    return -1;
  }
uint8_t get_energy_meter[] = { 0x01, 0x0a, 0x00, 0x13, 0x03, 0x03,
0x32, 0x01, 0x00, 0x25, 0x16, 0xe6 };
  //Send Energy Meter Value
  write_to_serial_device(get_energy_meter, 12);
if (pthread_join(reader, NULL)) {
    fprintf(stderr, "Error joining thread\n");
    return 2;
  }
  close_serial_device(device_file);
  return EXIT_SUCCESS;
}
```

In this sample, we have a Z-Wave command to send to the corresponding node and Fibaro wall plug to receive energy readings as raw values. In this command, the first four bytes form the home ID, the fifth byte forms the node ID or device ID assigned by the controller, then we have the command code and address, and finally the checksum of the bytes.

When we build and run the application, we will have the following outputs. The first received byte is an acknowledgment and then the next 18 bytes are for data.

```
root@clanton:~# ./zwave-galileo

Writing: 0x1 0xa 0x0 0x13 0x3 0x3 0x32 0x1 0x0 0x25 0x16 0xe6

Write Successful

Received 1 Bytes: 0x6

Received 18 Bytes: 0x1 0x10 0x0 0x4 0x0 0x3 0xa 0x32 0x2 0x21 0x44
0x0 0x0 0x0 0x33 0x0 0x0 0x84
```

Communication with remote Z-Wave devices is being handled with similar messages sent to the controller. For example, in order to query the controller for the home ID and device ID, you can send "0x1 0x3 0x0 0x20 0xdc" bytes to get 10 bytes of data, four bytes (the fifth to eighth bytes), the home ID, and the ninth byte to represent the node ID of the controller. Some common messages used to get data from the serial controller are listed in the following table:

Message sent	Message task
0x1 0x3 0x0 0x20 0xdc	Requests the home ID and node ID from Z-Wave controller.
0x01, 0x03, 0x00, 0x15, 0xe9	Requests the Z-Wave version of the serial controller.
0x01, 0x03, 0x00, 0x07, 0xfb	Requests the capabilities of the serial controller. The serial API version, manufacturer ID, product type, and product ID can be learned from this message.
0x01, 0x03, 0x00, 0x02, 0xfe	Requests initial data from the serial controller. Responses includes information about devices included in the device.

You can follow up with Z-Wave messages (frames) with open source projects for more information on this.

The following wiki page shows the Z-wave command classes: http://wiki.micasaverde.com/index.php/ZWave_Command_Classes

Summary

During the course of this chapter, we have tried to develop two simple applications. First, we have gone over all the C application development steps for a temperature sensor and talked about the importance of temperature data in home automation. We also measured relative humidity from the sensor, which it already calculates.

Then we discussed remote switches and tried to work on a device that exists in the market. In order to work on this remote switch, we included a controller sensor to communicate via the serial protocol with Intel Galileo. In order to make our application work with serial devices, we learned about building kernel modules for new devices attached to Intel Galileo.

Finally, we implemented functions to help us send and receive data from the serial Z-Wave controller and tried our samples to control energy consumption values from the remote wall plug.

In the next chapter, we will cover more on remote switches, that control light bulbs and lamp holders, and get more information on handling basic Z-Wave messages. While covering new devices, we will make our application more automated by adding threading capabilities.

4

Energy Management with Light Sensors and Extending Use Cases

We will follow on from the previous chapter with some new home automation devices and sensors, which are usable with Intel Galileo, and are related to energy management. In this chapter, we will try to focus on the lighting of our home. We'll check out some existing devices which you can use with Intel Galileo.

While finalizing the previous chapter, we made an introduction to the process of sending and receiving commands to Z-Wave devices with a USB controller; here, we will get into more detail about the use of Z-Wave commands.

Using light sensors

Light sensors are used to measure the current intensity of light in the ambient environment. Light intensity is measured in lumen units depending on the sensor and the amount of light emitted by the source. Lux is also used to indicate the light amount in the environment Lux means the luminous flux per unit area, which is equal to one lumen per square meter. For example, when there is full daylight, the amount of light is around 10000-25000 Lux. During full moon days, the amount of light is around 0.267 Lux.

Light intensity data can help you to automate your lighting system to switch it on or off. Switching off light automatically with your home automation system, where light intensity is high enough to see, will save energy. You can also create your own luminance data table and use it to automate your application to regulate when to switch lights on or off.

There are many sensors compatible with the Intel Galileo pinout. You can either use photocells or digital light sensors to make your circuit with Intel Galileo and implement your application to measure the illumination of the environment.

The following figure shows a photocell. Photocells act as resistors when the current light intensity that increases resistance would decrease on the photocell and you would start to read high voltages from the Intel Galileo analog pins.

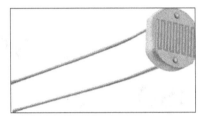

The next figure shows a digital light sensor manufactured by Adafruit. Adafruit TSL2561 uses the I2C protocol to communicate with Intel Galileo. In order to get the luminance value of your environment, you need to connect the sensor to Intel Galileo's corresponding pins for I2C and start reading.

 More information about the Adafruit sensor can be found at the following link: http://www.adafruit.com/product/439

You can find many other sensors available to use with Intel Galileo. Lighting sensors provide you with useful data to save energy to automate your lighting system.

Smart light bulbs

There is an increasing trend of using new wireless light bulbs for home appliances. There are many light bulbs available that use Zigbee, Z-Wave, Bluetooth 4.0, and other home automation protocols. Some of the light bulbs only allow remote switching on and switching off; some allow remote dimming of the light intensity of the bulb. New LED Light Bulbs also allow you to set the color of the light bulb by sending the RGB value to the light bulbs.

Using Philips Hue with Intel Galileo

One of the most popular wireless light bulb systems is Philips Hue. The Philips Hue system includes a bridge and remote led bulbs. The Philips Hue bridge connects with a home network to allow the user to manage light bulbs via a local network connection. Philips Hue bulbs use the Zigbee Light Link protocol, and the Philips Hue bridge controls the bulbs.

Philips provides software development kits for users to access the bridge and control light bulbs with REST API. Official SDKs and tools are provided for iOS, OS X, and Java platforms, but third-party developers ported SDK for many other platforms such as Node.js, ActionScript, Bash Script, Qt, C++, C#, Go, Perl, PHP, Python, and so on.

 You can access the Philips Hue developer website at the following link: `http://www.developers.meethue.com/`.

The Intel Galileo Linux image that we have created does support Node.js, Python, Bash Scripts, and so you can pick any of them to start developing an application for Philips Hue on your Intel Galileo.

Extending a home automation system with lighting control

We looked at lighting sensors and devices used for home automation. Let's go back to our home automation application to extend its use cases by connecting our Z-Wave wall plug to a desk lamp and using a Z-Wave lamp holder to read illumination values from a remote Z-Wave sensor.

In the previous chapter, we just made a brief introduction to using Z-Wave devices with a Z-Wave USB controller. During the course of this chapter, we will learn more about Z-Wave commands and how to control lighting devices from our Z-Wave USB controller.

The following picture shows our new device, a Z-Wave lamp holder produced by Everspring. We have included it in the Z-Wave network, as defined in the previous chapter, and we will add the ability to switch on and switch off the device remotely from Intel Galileo.

Our next circuit is a desk lamp attached to a Z-Wave wall plug from Fibaro. We have already included the device in our Z-Wave network. We will add the ability to switch on and off the desk lamp remotely from our application.

The last device in the following image includes multiple sensors produced by Philio which are a temperature sensor, illumination sensor, door/window sensor, and a motion sensor. In this chapter, we will try to only obtain illumination from a sensor. In the following chapter, we will use the motion and door/window sensor ability for security appliances.

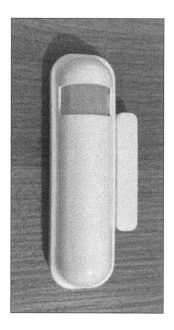

Continuing to home automation application development

In the previous chapter, we developed a simple application to read temperature and humidity values from an SHT11 sensor. We will use the functions from that application in our new application. You can start a new project from Eclipse or edit the previous ones.

Before proceeding, let's create new sources to increase the code reuse if needed. First, let's create our own GPIO library for basic functionality. In order to do that, we have added the galileo_gpio.h and galileo_gpio.c files to our project and copied our previously implemented functions into the galileo_gpio.c file; we also copied the function signatures to the galileo_gpio.h file and our header file looked like this:

```
#ifndef GALILEO_GPIO_H_
#define GALILEO_GPIO_H_
//GPIO VALUES
#define HIGH 1
#define LOW 0
#define INPUT "in"
#define OUTPUT "out"
//GPIO BUFFER
#define BUF 8
#define MAX_BUF 256
int gpio_set_value(int gpio_num, int value);
int gpio_set_mode(int gpio_num, const char* mode);
int gpio_export(int gpio_num);
int gpio_get_value(int gpio_num);
#endif /* GALILEO_GPIO_H_ */
```

This will enable us to use these functions for any new interaction with sensors previously mentioned and reuse the code. You can create or export any other libraries for your use to your project folder.

Then we created the thermometer.c source and the thermometer.h files to our project to define the thermometer code as a separate module. The defined function signatures in the thermometer.h file and header file will look like this:

```
#ifndef THERMOMETER_H_
#define THERMOMETER_H_
#include <stdint.h>
#define LSBFIRST 0
#define MSBFIRST 1
#define DATA_PIN 38  //IO 7
#define CLOCK_PIN 40 //IO 8
int shiftIn(int bit_order, int n_bits);
```

```
void shiftOut(uint8_t bitOrder, uint8_t val);
void delay(unsigned long ms);
void delayMicroseconds(unsigned int us);
float read_humidity();
float read_temperature();
void skip_crc();
#endif /* THERMOMETER_H_ */
```

And finally, we will create `serial.c` and `serial.h` files for serial communication with our Z-Wave USB adapter. We have made some little changes to our function signatures to make them reusable for other classes in our project. The preceding changes will give us the ability to make a more modular design and add many more features easily. The `serial.h` header file will be like this:

```
#ifndef SERIAL_H_
#define SERIAL_H_
#include <stdint.h>
int open_serial_device(const char* serial_device_path);
int close_serial_device(int device_file);
int read_from_serial_device(int device_file, uint8_t* data);
int write_to_serial_device(int device_file, uint8_t buffer[], int
  length);
#endif /* SERIAL_H_ */
```

Following the application, we will use very basic Z-Wave messages to handle messages from devices and control them. In order to interact with the device, we will implement a very basic terminal user interface, as follows, to read from stdin to execute the command. As we described in the previous section, we still keep our reader thread to handle incoming messages continuously. Our simple user interface is shown here. We will develop better interfaces in later chapters.

```
root@clanton:~/apps# ./smart_home

Successfully Set the Serial Device Parameters

Starting Home Manager

1 : Show Current Temperature

2 : Show Current Relative Humidity

3 : Get Current Status of Lamp Holder

4 : Switch On Lamp Holder

5 : Switch Off Lamp Holder

6 : Get Current Status of Wall Plug

7 : Switch On Wall Plug

8 : Switch Off Wall Plug

9 : Get Current Power Consumption
```

```
10: Get Energy Consumption

11: Get Power Level of Wall Plug

12: Get Power Level of Lamp Holder

13: Get Luminance of Environment

14: Quit

Enter Command to Execute:
```

The preceding command is our smart home application's user interface. The first two commands will call read_temperature and read_humidity functions from the first sample application developed with the SHT11 sensor, and so the SHT11 sensor should be connected to Intel Galileo. Other commands execute a simple command to manage remote Z-Wave devices.

Understanding Z-Wave commands

In the previous chapter, we executed a command but didn't look into any details. The message in Z-Wave can be more complex, but we will only look into the basic ones, which will help us to switch the wall plug and lamp holder on and off.

Z-Wave commands are a list of bytes, each of which indicates a specific value for a message type. Z-Wave messages read or sent to a serial controller start with 0x01, 0x06, 0x15, and 0x18. If the first value is 0x01, it shows the start of the message. If the received message from the serial controller is 0x06, it is an acknowledgment that the message is received. If it is 0x18, it means the message can't send and so it notifies you to resend a message. Let's continue with the following simple *Switch ON* message in the table.

An example of *Switch OFF* message for Lamp Holder device to send a Z-Wave serial controller is this: 0x01, 0x0A, 0x00, 0x13, 0x04, 0x03, 0x25, 0x01, 0x00, 0x25, 0x01, 0x00.

Message Byte	Byte Translation
0x01	This indicates the start of the Z-Wave message.
0x0A	In this message, there are 10 more bytes after the second byte; this is the message length after the second byte.
0x00	This byte defines the type of message, and it is a request. If the message is responded to, it would be 1.
0x13	This defines the function of this message, which means we want to send data to the corresponding device. 0x13 is a constant value.
0x04	This byte shows the corresponding device's ID. The lamp holder ID is 4.

Message Byte	Byte Translation
0x03	This byte shows the data length. The following three bytes define the operation or requested action from the device. This byte would be bigger if we wanted to set temperature value on a thermostat; in this case, we only send binary data to switch the lamp holder on or off.
0x25	This defines the command type. 0x25 corresponds to the binary switch operation. Other commands are defined later in the section.
0x01	This shows whether we will set the value of switch or get the value of switch. 1 indicates that we will set the switch. If it is 0, the device will respond with the status of the switch and the following byte will not be in this message.
0x00	This byte indicates what value we want to set for the switch. 0 means that we will switch it off; 1 means we'll switch it on.
0x25	This byte indicates the transmit type to the corresponding device. 0x25 indicates to send a message to the indicated node directly.
0x01	This byte is the message identification assigned by the user or program; it is the callback ID of the message.
0x00	This byte should be generated later; this is a checksum value. In the following section, the calculation of the checksum value will be shown.

The first byte shows the start of the message as we've already described. The second byte is the number of bytes in the rest of the message. 0x0A means that there are 10 more bytes following this.

The third byte represents the message type, which means either *respond* or *request*. If it is a request, we place 0x00, and if it is a response, it is 0x01. The fourth byte represents a message function; 0x13 is the value for sending data to the serial controller.

The fifth byte in our example shows the node ID number in the Z-Wave controller. The Z-Wave controller automatically increases the node ID number when devices are included. In our example in this book, we first added the Philip multisensor, and so 0x02 is the multisensor device. 0x03 is the wall plug and 0x04 is the lamp holder. 0x01 is assigned to the Aeon Z-Wave controller.

After the node ID byte, 0x03 shows the length of the command to be executed by the remote device. So following 0x03 after the node ID, the three bytes are the command, which will be executed by the remote device.

After the command length information, the seventh byte shows the command class, which defines the type of action that the device needs to take and the type of value we have received. The following table shows some simple command classes and their values, which we will use in our example:

Command classes	Values	Description
COMMAND_CLASS_BASIC	0x20	Set and get basic data
COMMAND_CLASS_SWITCH_ALL	0x27	Switch on and off
COMMAND_CLASS_SWITCH_BINARY	0x25	Binary switch on and off
COMMAND_CLASS_METER	0x32	Receive or request meter values
COMMAND_CLASS_ALARM	0x71	Alarm data to broadcast
COMMAND_CLASS_POWERLEVEL	0x73	Power level of device
COMMAND_CLASS_BATTERY	0x80	Battery level of device
COMMAND_CLASS_SENSOR_BINARY	0x30	Binary sensor like motion sensor, if there is motion or not
COMMAND_CLASS_SENSOR_MULTILEVEL	0x31	Sensor representing more than a binary value like a temperature sensor
COMMAND_CLASS_SENSOR_MULTILEVEL_V2	0x31	Similar to the previous command class
COMMAND_CLASS_SENSOR_ALARM	0x9C	When sensor raises an alarm

These commands classes can only be used in the devices which support them. If you send a node a command including battery and if it doesn't have a battery, there won't be any response. Basically, commands are usually used to get the status of the device if the latter is online and in the network. Almost all device manuals include information about supported command classes for the device, and so you can check your device's manual to see what command classes you can use it with.

More details and more command classes can be learned from open source projects such as OpenZWave, LinuxMCE, or the RaZberry project. The following links are great reading material to get more information about Z-Wave commands:

- `https://github.com/yepher/RaZBerry`
- `http://wiki.micasaverde.com/index.php/ZWave_Command_Classes`
- `http://wiki.zwaveeurope.com/index.php?title=Z-Wave_Application_Layer`

You can also check Z-Wave SDK from Z-Wave Alliance for more information at the following link:

`http://www.z-wavealliance.org/z-wave-for-developers-oems`

The following command class value is in the message; there is a byte for the operation type if it is for setting a value in the remote device or requesting a report. 0x01 defines a setting operation and 0x02 is for getting information from a device. As we want to switch off the device, we need to send the switch off value. The switch off value is 0x00, and the switch on value is 0xFF for binary switches.

Finally, there is a transmitting type value 0x25 defined for Aeon Stick. It follows with a callback ID, which you can check from the received messages, that the message has been received by the device. Finally, the checksum value is appended in the message buffer. All messages have a checksum value at the last byte.

It will be clearer while we proceed on our example coding while trying to manage the appliances at home from Intel Galileo.

Handling Z-Wave messages from Intel Galileo

In order to send and receive Z-Wave messages to/from Z-Wave USB adapter, message classes have been defined. The message.c and message.h files are created to store functions that create messages to send and parse the incoming messages. We have also defined the necessary values to be used in the function in the message.h header file, which are shown next:

```
#ifndef MESSAGE_H_
#define MESSAGE_H_
#include <stddef.h>
#include <stdint.h>
#define BUFFER 256
#define REQUEST    0x00
#define RESPONSE    0x01
#define BASIC_SET  0x01
#define BASIC_REPORT  0x03
#define COMMAND_CLASS_CONTROLLER_REPLICATION    0x21
#define COMMAND_CLASS_APPLICATION_STATUS    0x22
#define COMMAND_CLASS_HAIL            0x82
#define TRANSMIT_OPTION        0x25
#define ControllerNodeID    0x01
#define MultiSensorNodeID    0x02
#define WallPlugNodeID    0x03
#define LampHolderNodeID    0x04
/*
 * Message Type
 */
typedef enum ZWAVE_MESSAGE_TYPE {
```

```
     SOF = 0x01, ACK = 0x06, NAK = 0x15, CAN = 0x18,
   } MESSAGE_TYPE;
/*
 * Message Function
 */
enum FUNCTION {
  SEND_DATA = 0x13, RESPONSE_RECEIVED = 0x04
};
/*
 * Multilevel Sensor Type
 */
enum SENSOR_TYPE {
  TEMPERATURE_SENSOR = 0x01, LUMINANCE_SENSOR = 0x03, POWER_SENSOR
  = 0x04
};
/*
 * Binary Sensor Value
 */
enum BINARY_SENSOR_VALUE {
  ON = 0xFF, OFF = 0x00,
};
/*
 * Sensor Message Types
 */
enum SENSOR_COMMANDS {
  BINARY_SET      = 0x01,
BINARY_GET      = 0x02,
BINARY_REPORT   = 0x03, // Response to the Get Command
  MULTILEVEL_GET    = 0x04,
  MULTILEVEL_REPORT   = 0x05
};
/*
 * Energy Meter Message Types
 */
enum ENERGY_METER_COMMANDS {
  METER_GET         = 0x01,
  METER_REPORT      = 0x02,
  METER_SUPPORTED_GET   = 0x03,
  METER_SUPPORTED_REPORT   = 0x04,
  METER_RESET       = 0x05
};
/*
 *WALL Plug Meter Type
 */
```

```
#define ENERGY      0x00
#define POWER       0x10
/*
 * ZWave Command Classes
 */
enum ZWAVE_COMMAND_CLASS {
  COMMAND_CLASS_NO_OPERATION      = 0x00,
  COMMAND_CLASS_BASIC         = 0x20,
  COMMAND_CLASS_SWITCH_ALL        = 0x27,
  COMMAND_CLASS_SWITCH_BINARY      = 0x25,
  COMMAND_CLASS_METER         = 0x32,
  COMMAND_CLASS_ALARM         = 0x71,
  COMMAND_CLASS_POWERLEVEL        = 0x73,
  COMMAND_CLASS_BATTERY          = 0x80,
  COMMAND_CLASS_SENSOR_BINARY       = 0x30,
  COMMAND_CLASS_SENSOR_MULTILEVEL    = 0x31,
  COMMAND_CLASS_SENSOR_MULTILEVEL_V2    = 0x31,
  COMMAND_CLASS_SENSOR_ALARM       = 0x9C
};
/*
 * Functions to Handle Messaging
 */
uint8_t generate_checksum(uint8_t buffer[], int length);
int parse_incoming_mesage(uint8_t* message, int length);
int handle_incoming_message(int serial_device, uint8_t message[],
  int length);
int binary_switch_on_off(int serial_device, uint8_t nodeID,
  uint8_t on_off,uint8_t callbackID);
int get_meter_level(int serial_device, uint8_t nodeID, uint8_t
  type,
    uint8_t callbackID);
int get_binary_switch_status(int serial_device, uint8_t nodeID,
    uint8_t callbackID);
int get_node_power_level(int serial_device, uint8_t nodeID,
  uint8_t callbackID);
int get_luminance_value(int serial_device, uint8_t nodeID, uint8_t
  sensor_type,uint8_t callbackID);
#endif /* MESSAGE_H_ */
```

We have defined all the necessary functions and constants for use in the application. In our main function, we will continue to use the reader thread as in the previous chapter, but we will make the thread also handle incoming messages to Intel Galileo.

```
void* reader_thread(void *arg) {
  while (1) {
    uint8_t data[256];
    int m_length = read_from_serial_device(device_file, data);
```

```
      if (m_length > 0) {
        handle_incoming_message(device_file, data, m_length);
      }
    }
  }
```

The commands will be handled with a switch that picks the right function to call
and execute a command. As shown in following code snippet, the main function
reads the user input and sends it to the execute_command function to start the
required function:

```
int main(int argc, char* argv[]) {
  //Open Serial Device
  device_file = open_serial_controller(serial_device_path);
  if (device_file < 0) {
    printf("Can't Open Serial Device %s", serial_device_path);
    return EXIT_FAILURE;
  }
  int err = pthread_create(&reader, NULL, &reader_thread, NULL);
  if (err != 0) {
    close_serial_controller(device_file);
    printf("Can't create Thread\n");
    return -1;
  }
  printf("Starting Home Manager\n");
  int choice = 0;
  user_menu();
  while (choice != 14) {
    printf("Enter Command to Execute:");
    scanf("%d", &choice);
    execute_command(choice);
  }
  if (pthread_join(reader, NULL)) {
    fprintf(stderr, "Error joining thread\n");
    return 2;
  }
  close_serial_device(device_file);
  return EXIT_SUCCESS;
}

void execute_command(int choice) {
  switch (choice) {
  case 1:
    printf("Current Temperature is: %f Celcius\n",
    read_temperature());
```

```
  break;
case 2:
  printf("Current Temperature is: %f Celcius\n",
read_humidity());
  break;
case 3:
  get_binary_switch_status(device_file, LampHolderNodeID, 1);
  break;
case 4:
  binary_switch_on_off(device_file, LampHolderNodeID, ON, 2);
  break;
case 5:
  binary_switch_on_off(device_file, LampHolderNodeID, OFF, 3);
  break;
case 6:
  get_binary_switch_status(device_file, WallPlugNodeID, 4);
  break;
case 7:
  binary_switch_on_off(device_file, WallPlugNodeID, ON, 5);
  break;
case 8:
  binary_switch_on_off(device_file, WallPlugNodeID, OFF, 6);
  break;
case 9:
  get_meter_level(device_file, WallPlugNodeID, POWER, 7);
  break;
case 10:
  get_meter_level(device_file, WallPlugNodeID, ENERGY, 8);
  break;
case 11:
  get_node_power_level(device_file, WallPlugNodeID, 9);
  break;
case 12:
  get_node_power_level(device_file, LampHolderNodeID, 10);
  break;
case 13:
  get_luminance_value(device_file, MultiSensorNodeID,
LUMINANCE_SENSOR, 11);
  break;
case 14:
  printf("Quitting.....");
  break;
default:
  break;
}}
```

Reading the status of remote devices from Intel Galileo

Let's start with requesting the current status of the wall plug and the lamp holder from Intel Galileo. When we request the status, the following function will be executed:

```
int get_binary_switch_status(int serial_device, uint8_t nodeID,
    uint8_t callbackID) {
    int message_length = 11;
    uint8_t checksum = 0x00;
    uint8_t message_buffer[] = { SOF, 0x09, REQUEST, SEND_DATA,
    nodeID, 0x02, COMMAND_CLASS_SWITCH_BINARY, BINARY_GET,
    TRANSMIT_OPTION, callbackID, checksum };
    checksum = generate_checksum(message_buffer, message_length);
    message_buffer[message_length - 1] = checksum;
    return write_to_serial_device(serial_device, message_buffer,
    message_length);
}
```

We send the serial device file, node ID, and callback ID parameter to the function and create the message with COMMAND_CLASS_SWITCH_BINARY and BINARY_GET to receive the status of the device. This message length is 11 bytes as we are not setting any value and just making a status request. The following is the output received when we request the status of the lamp holder:

```
Enter Command to Execute: 3
Writing: 0x1 0x9 0x0 0x13 0x4 0x2 0x25 0x2 0x25 0x1 0xe0
Write Successful
Enter Command to Execute:Received: 0x6
ACK Received
Writing: 0x6
Write Successful
Received: 0x1 0x4 0x1 0x13 0x1 0xe8
Data Sent to ZWave Stack
Received: 0x1 0x5 0x0 0x13 0x1 0x0 0xe8
Data Request with Callback ID 0x1 Received
Received: 0x1 0x9 0x0 0x4 0x0 0x4 0x3 0x25 0x3 0xff 0x2c
Response From Lamp Holder Node Received: Status of Device is ON
```

Let's parse the received message. The first byte is 0x1, which is the start of the message. 0x9 is the length of the rest. The Node ID is the sixth byte, which is the lamp holder node ID. The next byte is 0x3, which shows the length of the command. If the 0x25 command class is received, we have 0x3, which means the report from the sensor. Finally, we have the 0xff value to indicate the status of the device, which is on here.

Switching the lamp holder on/off with Intel Galileo

Let's switch off the lamp holder and then check its status. We will execute the function given below. The following message is similar to the one we examined in the previous chapter:

```
int binary_switch_on_off(int serial_device, uint8_t nodeID,
  uint8_t on_off,
    uint8_t callbackID) {
  int message_length = 12;
  uint8_t checksum = 0x00;
  uint8_t message_buffer[] = { SOF, 0x0a, REQUEST, SEND_DATA,
  nodeID, 0x03,
      COMMAND_CLASS_SWITCH_BINARY, BINARY_SET, on_off,
  TRANSMIT_OPTION, callbackID, checksum };
  checksum = generate_checksum(message_buffer, message_length);
  message_buffer[message_length - 1] = checksum;
  return write_to_serial_device(serial_device, message_buffer,
  message_length);
}
```

When we execute the command, we need to send this function the node ID, the on or off value at that time, and the callback ID. The execution is shown next:

```
Enter Command to Execute: 5
Writing: 0x1 0xa 0x0 0x13 0x4 0x3 0x25 0x1 0x0 0x25 0x3 0xe3
Write Successful
Enter Command to Execute: Received: 0x6
ACK Received
Writing: 0x6
Write Successful
Received: 0x1 0x4 0x1 0x13 0x1 0xe8
Data Sent to ZWave Stack
Received: 0x1 0x5 0x0 0x13 0x3 0x0 0xea
Data Request with Callback ID 0x3 Received
Enter Command to Execute:3
Writing: 0x1 0x9 0x0 0x13 0x4 0x2 0x25 0x2 0x25 0x1 0xe0
Write Successful
Enter Command to Execute: Received: 0x6
ACK Received
Writing: 0x6
Write Successful
Received: 0x1 0x4 0x1 0x13 0x1 0xe8
Data Sent to ZWave Stack
```

```
Received: 0x1 0x5 0x0 0x13 0x1 0x0 0xe8
Data Request with Callback ID 0x1 Received
Received: 0x1 0x9 0x0 0x4 0x0 0x4 0x3 0x25 0x3 0x0 0xd3
Response From Lamp Holder Node Received: Status of Device is OFF
```

After we execute the command, we get ACK from the controller and command callback ID, which shows that our message transmitted successfully.

Handling incoming messages

As we've seen in the terminal output, we have parsed the incoming messages. It is not easy to handle messages, but we need to check each command class to decide what value has been received to the Z-Wave controller.

We can simply identify the fourth byte for what type of value is received. If it is 0x04, we know that a sensor has sent a message or response. Then we know that we need to check the sixth byte to check the node ID. It follows with the length of the command, command class, length of value, and the sensor or device value. A very basic parsing of a received message from the binary switch or a multilevel sensor is given here:

```
uint8_t length_of_rest = message[1];
uint8_t message_type = message[2];
uint8_t message_function = message[3];
uint8_t data_length = message[6];
uint8_t command_class = message[7];
if (message_function == RESPONSE_RECEIVED) {
  printf("Response From ");
  if (message[5] == MultiSensorNodeID) {
    printf("Multi Sensor Node Received: ");
  } else if (message[5] == WallPlugNodeID) {
    printf("Wall Plug Node Received: ");
  } else if (message[5] == LampHolderNodeID) {
    printf("Lamp Holder Node Received: ");
} }
if (command_class == COMMAND_CLASS_SWITCH_ALL) {
printf("Status of Device is ");
  if (message[9] == OFF) {
    printf("OFF\n");
  } else if (message[9] == ON) {
    printf("ON\n");
} else if (command_class == COMMAND_CLASS_SENSOR_MULTILEVEL_V2) {
    if (message[8] == MULTILEVEL_REPORT) {
      if (message[9] == TEMPERATURE_SENSOR) {
```

```
    printf("Temperature Value is %d Fahreneit\n", message[12]);
} else if (message[9] == LUMINANCE_SENSOR) {
    printf("Illumination is %d % \n", message[11]);}}}
```

The Philio multisensor reports when there is a change in any of the sensors. Let's hold our desk lamp directly to the Philio multisensor to measure luminance and then let's switch off the wall plug with command 8. Now, read the luminance value change:

```
Received: 0x1 0xb 0x0 0x4 0x0 0x2 0x5 0x31 0x5 0x3 0x1 0x5a 0x9b
Response From Multi Sensor Node Received: Illumination is 90%
Enter Command to Execute: 8
Received: 0x6
ACK Received
Writing: 0x6
Write Successful
Received: 0x1 0x4 0x1 0x13 0x1 0xe8
Data Sent to ZWave Stack
Received: 0x1 0x5 0x0 0x13 0x6 0x0 0xef
Data Request with Callback ID 0x6 Received
Received: 0x1 0xb 0x0 0x4 0x0 0x2 0x5 0x31 0x5 0x3 0x1 0x5 0xc4
Response From Multi Sensor Node Received: Illumination is 5%
```

These kinds of sensor data are very useful to automate your home. You can add the logic to your application to switch off the light if the illumination received is more than 20 percent or switch it on if it is less than 5 percent according to this sensor.

Summary

In this chapter, we tried to cover devices and sensors related to lighting and light intensity measurement and their use in the Linux application running on Intel Galileo.

In the first part, we looked into the sensors compatible with Intel Galileo, which may help you obtain light intensity measurement and use the data for your home. Then we followed up with remote light bulbs and mentioned the popular ones such as Philips Hue and its SDKs, which are compatible with Intel Galileo.

Remote light bulbs are getting more popular by the day with other protocols, especially with the Zigbee Light Link protocol. With the information given in this chapter, you should have an idea on how to automate your lighting system. We set up some basic circuits to show the usage. We connected our remote wall plug to a desk lamp and used a Z-Wave lamp holder for the room bulb. You can extend these cases by adding different devices and use remote wall plugs or bulbs as you need.

Along with this chapter, we also finished looking at the energy management concept, and we will follow up with the security concept of home management and will also talk about related sensors and devices as we did in *Chapter 2, Getting Started with Home Automation Applications* and *Chapter 3, Energy Management with Environmental and Electrical Sensors.*

In the following chapter, we will start learning about the security concept in home automation. We will discover new devices and sensor use cases of our home automation application. We will also continue to iterate our smart home application with one more step by adding new features.

5

Home Monitoring with Common Security Sensors

In *Chapter 3*, *Energy Management with Environmental and Electrical Sensors* and *Chapter 4*, *Energy Management with Light Sensors and Extending Use Cases*, we worked with sensors, to achieve energy management. Now, we will proceed to learn about the security concept in home automation and use cases with Intel Galileo.

Security is another major topic in home automation. Securing a residential area while creating a home automation system is done with various sensors and devices. For example, a motion sensor can be used to monitor an area if someone breaks into it. A door or a window sensor can be used for a similar purpose such as monitoring a door and window to check whether they have been opened without your knowledge. Network cameras are another set of devices used for security in home automation systems to monitor homes or any other area included remotely.

Besides securing residential areas from burglary, you must also look out for devices that sense fire, gas leaks, or water leaks. Smoke detectors, gas detector, water leaks, and flood detectors can be included into the home automation system. With the addition of these sensors, any damage or disaster can be prevented.

In this chapter, we will briefly examine the sensors existing in the market, that are usable with Intel Galileo for security. We will also look at devices with home automation protocols.

Security sensors with Intel Galileo

There are many sensors, that can be used with Intel Galileo to create security devices. Let's look at some of them.

PIR motion sensors

There are many motion or **passive infrared detection (PIR)** sensors available to be used with Intel Galileo. The sensor shown in the following image, manufactured by SeeedStudio, is one of them. Their use is pretty straightforward. All you need to do is connect VCC, Ground, and one GPIO pin for detection output. After you have connected pins, you will read *HIGH (1)* from the connected GPIO pin; otherwise, you would read *LOW (0)*.

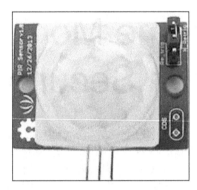

 You can find more information about motion sensors from SeeedStudio from the following link: http://www.seeedstudio. com/wiki/PIR_Motion_sensor_module

Motion sensors can be used to develop a device that can sense movement and alert the user. Another use case with energy management is that you can turn on or off a light bulb after a motion has been detected.

Magnetic sensors

These sensors are the main mechanism used for door/window sensors. Digital magnetic sensors work at a very basic level. They produce a digital high value if there is a magnetic object near the sensor. The following image shows a very basic sensor from DFRobot, whose pins can be connected to VCC, GND, and GPIO to read digital values in order to sense whether any magnetic object is nearby:

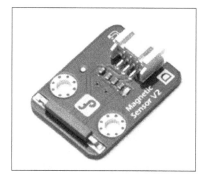

The preceding image shows a magnetic sensor from dfrobot; more information about this can be found at the following link `http://www.dfrobot.com/wiki/index.php/Digital_magnetic_sensor_SKU:_DFR0033`.

By using a magnetic sensor, it is possible to develop a device to detect if a door or window has been opened. It is also possible to use magnetic sensors for energy management; if any door has been opened, you can switch on the lights in the given room.

Gas sensors

There are many gas sensors available to be used with Intel Galileo. Each gas sensor is able to detect a specific gas. Some sensors detect the levels of carbon monoxide, methane, propane, and alcohol in the air. It helps you to detect any rise in the level of dangerous gases so that you can take precautions immediately to prevent any accident.

Using gas sensors with Intel Galileo

Here, we have an example of a **MQ-9** gas sensor, which is sensitive to carbon monoxide, methane, and liquid petrol gas. Changes in the level of the given gases affect the conductivity of the sensor. A change in conductivity results in the output voltage of the sensor increasing. MQ-9 is an analog sensor, and so we need to use analog pins in Intel Galileo to read voltage changes.

There are multiple sources for similar MQ-9 datasheets. Just search using your favorite search engine. One of the sources is here: `http://www.sgbotic.com/products/datasheets/sensors/MQ-9.pdf`

The following image is the MQ-9 sensor we've used in the following sample application:

1. First, we need to set the Intel Galileo GPIO pin to 37 low to be able to read analog values from analog input 0. Our `main` function will look like this:

```
#define PINMUX 37
#define ANALOGPIN 0
/**
 * Analog Device File Operations
 */
int open_analog_device(int pin_number);
int read_analog_device_value(int device_file);
float read_voltage_scale(int pin_number);
int main(void) {
  // Read from Analog 0, mux GPIO Pin 37 to read voltage
  values
  if (gpio_set_mode(PINMUX, OUTPUT) < 0) {
    printf("Can't Set GPIO Mux Pin Mode\n");
    return EXIT_FAILURE;
  }
  if (gpio_set_value(PINMUX, LOW) < 0) {
    printf("Can't Set GPIO Mux Pin Value\n");
    return EXIT_FAILURE;
  }
  printf("Pin Mux Successful\n");
  // Read Digital Value Scale
  float v_scale = read_voltage_scale(ANALOGPIN);
  printf("Voltage Scale is %f\n", v_scale);
  // Open Analog Device
  int analog = open_analog_device(ANALOGPIN);
  if (analog > -1) {
    printf("Analog IO File Opened Successfully\n");
```

```
    }
     // Read Voltage Values from Analog 0
    while (1) {
      printf("Voltage : %d \n", read_analog_device_value(analog));
      usleep(1000 * 1000);
    }
    close(analog);
    return 0;
}
```

2. Then, we will read the raw analog values from the filesystem. We need to open /sys/bus/iio/devices/iio\:device0/in_voltage0_raw and read the values:

```
int open_analog_device(int pin_number) {
  //Analog Device Values Read from
  const char* analog_file_path =
      "/sys/bus/iio/devices/iio\:device0/in_voltage%d_raw";
  int fanalog, a_err = -1;
  char analog_file_buffer[MAX_BUF];
  //Set analog reading file path and pin number
  if (sprintf(analog_file_buffer, analog_file_path,
  pin_number) < 0) {
    printf("Can't create analog file path\n");
    return a_err;
  }
  fanalog = open(analog_file_buffer, O_RDONLY);
  if (fanalog < 0) {
    printf("Can't open analog device\n");
    return a_err;
  }
  return fanalog;
}
int read_analog_device_value(int device_file) {
  char buf[1024];
  int a_err = -1;
  int ret = read(device_file, buf, sizeof(buf));
  if (ret == a_err) {
    printf("Couldn't get value from Analog Pin\n");
    return -1;
  }
  buf[ret] = '\0';
  lseek(device_file, 0, 0);
  return atoi(buf);
}
```

3. There is an extra piece of code to read the scale of the raw values for future conversion requirements. It is done by reading `/sys/bus/iio/devices/iio\:device0/in_voltage%d_scale`:

```
float read_voltage_scale(int pin_number) {
  //Analog Device Values Read from
  const char* scale_file_path =
      "/sys/bus/iio/devices/iio\:device0/in_voltage%d_scale";
  int fscale, a_err = -1;
  char scale_file_buffer[MAX_BUF];
  //Set analog reading file path and pin number
  if (sprintf(scale_file_buffer, scale_file_path,
  pin_number) < 0)
{
    printf("Can't create scale file path\n");
    return a_err;
  }
  fscale = open(scale_file_buffer, O_RDONLY);
  if (fscale < 0) {
    printf("Can't open scale device\n");
    return a_err;
  }
  char buf[1024];
  int ret = read(fscale, buf, sizeof(buf));
  if (ret == a_err) {
    printf("Couldn't get value from Analog Pin\n");
    return -1;
  }
  buf[ret] = '\0';
  return atof(buf);
}
```

4. Let's run the application on the Intel Galileo. We get the following output from our `gasdetector` application:

```
root@clanton:~/apps# ./gasdetector
Pin Mux Successful
Voltage Scale is 1.220703
Analog IO File Opened Successfully
Voltage : 652
Voltage : 748
Voltage : 776
Voltage : 2916
Voltage : 2848
Voltage : 2812
Voltage : 2728
Voltage : 2708
```

This was a basic sample working with analog sensors. Gas sensors also have a very important place in security. The reliable use of gas sensors can prevent many accidents in residential and industrial areas.

Security devices for home automation

We have looked at some of the sensors available for use with Intel Galileo. In this section, we will explore some of the devices already in the market and ready to use with home automation systems.

Motion, window, and door lock sensors

For security, there are many devices out in the market with most of the home automation protocols. Let's first look at door/window sensors and motion sensors. There are door/window sensors with almost all of the home automation protocols. For example, we have the Philio multi-sensor device, which includes a door/window sensor with a magnetic piece to detect whether the door or window is open or closed. You can find the door/window sensor sold separately with your selected home automation protocol to add to your home automation system.

The following image shows the Z-Wave Philio multi-sensor. It is for when door or a window is open. In a home automation system, you can get multiple door/window sensors in your home and can customize the names of the sensors to understand which room's door or window is open at a given time.

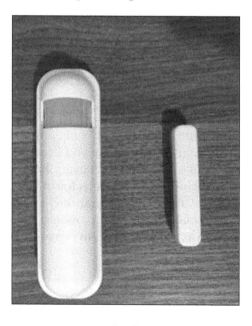

As shown in the preceding image, on the Philio multi-sensor device, you will also see a motion sensor at the head of the bigger part of the device. Motion sensors can also be used separately and you can place them in various places to detect motion.

Besides the door/window and motion detectors, you can also find door locks, that replace classical door locks. These locks allow you to remotely monitor and control the lock system. For example, you can connect to your home automation system remotely to check if the door is locked and, if it's not, you can lock it remotely.

> The following image shows one of the door locks that can be used with Z-Wave and ZigBee by Yale. More information about the product can be found at `http://www.yaleresidential.com/en/yale/yaleresidential-com/Residential/?groupId=1294297&productId=1294301`.

Smoke and flood detectors

Preventing dangerous events from occurring or being notified when they are happening is key to securing your home. Smoke and flood detectors will help you to do that. There are many smoke detectors with every home automation protocol for fire detection. When you include a smoke detector in your home automation system, the home automation system can automatically switch the wall plugs off immediately to prevent fire damage to electricity outlets. Any other action can be defined according to the home automation system's capabilities.

There are also flood detectors available to alarm systems to detect any water leakage in the residential area. These can be placed onto your kitchen floor or anywhere in the residential area, that is ideally supposed to be dry. Flood detectors are useful when there is a broken pipe that leaks water or an amount of water has somehow entered an area that should be dry. Flood detectors might not help much during sudden natural disasters but will help you to detect water in normal situations.

In the following sections, we will use a Z-Wave flood detector and include it into our Smart Home application. We will use a flood detector (as shown in the following image) produced by Everspring. In the application, we will get the sensor status and take action according to it. We will make a test to close the lamp holder and wall plug if flood is detected by the sensor.

Let's start by adding new features to our Smart Home application that we have started in the previous chapters.

Adding security features to the Smart Home application

We have already added the Philio multisensor to our system to read illumination and temperature values. Let's add a flood detector to the Aeon USB controller; and as you have in previous chapters, follow inclusion instructions from its manual. Then we can start the required implementation in the application.

Before proceeding to work with security sensors, we need to define sensor types and command classes to identify messages.

The following commands are used to report sensor changes:

Sensor type	Message sent	Report value
Binary sensor	COMMAND_CLASS_BINARY_SENSOR, 0x30	BINARY_REPORT, 0x03
Sensor alarm	COMMAND_CLASS_SENSOR_ALARM, 0x9C	SENSOR_ALARM_ REPORT, 0x02

These constants have been defined in the message.h file, as shown here:

```
#define FloodSensorNodeID      0x05

enum BINARY_SENSOR_TYPE {
   GENERAL_PURPOSE = 0x01,
   WATER_DETECTION_SENSOR = 0x06,
   TAMPER_SENSOR = 0x08,
   DOOR_WINDOW_SENSOR = 0x0A,
   MOTION_DETECTION_SENSOR = 0x0C,
   GLASS_BREAK = 0x0D
};
/**
 * Sensor Alarm Types
 */
enum ALARM_TYPE {
   GENERAL_ALARM = 0x01,
   SMOKE_ALARM = 0x02,
   CARBON_MOXIDE_ALARM = 0x03,
   HEAT_ALARM = 0x04,
   FLOOD_ALARM = 0x05
};
/**
 * ALARM COMMANDS
 */
enum SENSOR_ALARM_COMMAND{
   SENSOR_ALARM_GET = 0x01,
   SENSOR_ALARM_REPORT = 0x02,
   SENSOR_ALARM_SUPPORTED_GET = 0x03,
   SENSOR_ALARM_SUPPORTED_REPORT = 0x04
};
.......... . .
/*
```

```
 * ZWave Command Classes
 */
enum ZWAVE_COMMAND_CLASS {

  .
  COMMAND_CLASS_ALARM = 0x71,
  COMMAND_CLASS_SENSOR_ALARM = 0x9C,
  COMMAND_CLASS_SILENCE_ALARM = 0x71,
  ......
  COMMAND_CLASS_SENSOR_BINARY = 0x30,
  ....... .
  COMMAND_CLASS_WAKE_UP = 0x84,
  ....... .
};
```

Motion detection

Motion sensors are binary sensors that report whether there is motion or not. We will make some additions to the message handler to identify the message from the motion sensor.

While we were dealing with the Z-Wave Temperature or Luminance sensor, we checked if the incoming message's command class is a multilevel sensor (0x31); this time we will check for the command class binary sensor (0x30) and check the sensor type. The following code needs to be added to the message handler to parse an incoming message from a motion sensor:

```
 ..
 else if (command_class == COMMAND_CLASS_SENSOR_BINARY) {

      if (message[8] == BINARY_REPORT) {

        if (data_length == 0x04) {


if (message[10] == MOTION_DETECTION_SENSOR) {

             if (message[9] == ON) {
               printf("Motion DETECTED\n");
             } else if (message[9] == OFF) {
               printf("NO Motion Detected\n");
}}}}}
```

When a motion is detected, the motion sensor sends a message to the Z-Wave controller and our Smart Home application receives and parses the message to create the following output:

```
Received: 0x1, 0xa, 0x0, 0x4, 0x0, 0x2, 0x4, 0x30, 0x3, 0xff, 0xc,
    0x37,
Response From Multi-Sensor Node Received: Motion DETECTED
```

Door/window sensor detection

Door/window sensors are binary sensors too. It reports when the door is opened, and so we will use the same command class as we did for the motion sensor. We have already added the sensor type in the previous section, and so we only need to add another case to control if the reporting binary sensor is a door/window sensor. Message handling for binary sensors will be as follows:

```
..
else if (command_class == COMMAND_CLASS_SENSOR_BINARY) {
    if (message[8] == BINARY_REPORT) {
        if (data_length == 0x04) {
          if (message[10] == DOOR_WINDOW_SENSOR) {
            if (message[9] == ON) {
              printf("Door/Window Sensor is OPEN\n");
            } else if (message[9] == OFF) {
              printf("Door/Window Sensor is CLOSE\n");
            }
          } else if (message[10] == MOTION_DETECTION_SENSOR) {
            if (message[9] == ON) {
              printf("Motion DETECTED\n");
            } else if (message[9] == OFF) {
              printf("NO Motion Detected\n");
            }
        } }}}
..
```

When a door/window sensor is opened, it will send a message to the Z-Wave controller. Our Smart Home application will parse the message and show the following output:

```
Received: 0x1, 0xa, 0x0, 0x4, 0x0, 0x2, 0x4, 0x30, 0x3, 0xff, 0xa,
    0x31,
Response From Multi-Sensor Node Received: Door/Window Sensor is
    OPEN
```

When the door/window sensor is closed, it will send a message indicating that the sensor status is closed.

```
Received: 0x1, 0xa, 0x0, 0x4, 0x0, 0x2, 0x4, 0x30, 0x3, 0x0, 0xa,
0xce,
Response From Multi-Sensor Node Received: Door/Window Sensor is CLOSE
```

Flood detection

Let's add our new device to the application. The Everspring Flood detector has a long cable ending with two metal pins, as seen in the previous image. When those two pins come into contact with water, they raise an alarm.

After inclusion of the flood detector, it will be assigned node ID 0x05 by the USB controller. The flood detector will send a sensor alarm to the home automation system. The flood detection alarm uses the COMMAND_CLASS_SENSOR_ALARM Z-Wave command class.

An alarm command class, alarm commands, and sensor types have been defined. We can proceed to handle incoming alarm messages from the flood detector. We will only try to deal with the sensor alarm at this moment.

In order to handle the sensor alarm, we need to handle a new command class case. The following lines will be added to our message handler to identify the sensor alarm and to alert the user. We also added the smoke alarm case here:

```
else if (command_class == COMMAND_CLASS_SENSOR_ALARM) {
  printf("Sensor Alarm ");
  if (message[8] == SENSOR_ALARM_REPORT) {
    printf("Reported ");
    if (message[10] == FLOOD_ALARM) {
      if (message[11] == ON) {
        printf("FLOOD Detected\n");
      } else if (message[11] == OFF) {
        printf("NO Flood Danger\n");
      }
    } else if (message[10] == SMOKE_ALARM) {
      if (message[11] == ON) {
        printf("SMOKE Detected\n");
      } else if (message[11] == OFF) {
        printf("NO Fire Danger\n");          }}}}
```

Let's make a test with the flood detector. We first put the metal pins into a glass of water to see if we are able to receive alarms from the flood detector.

A sample output from the Smart Home application is as follows:

```
Received: 0x1, 0xd, 0x0, 0x4, 0x0, 0x5, 0x7, 0x9c, 0x2, 0x0, 0x5,
  0xff, 0x0, 0x0, 0x90,
Response From Flood Sensor Node Received: Sensor Alarm Reported
  FLOOD Detected
```

When we take the pins out of the glass, the flood detector reports no flood detected to the application.

```
Received: 0x1, 0xd, 0x0, 0x4, 0x0, 0x5, 0x7, 0x9c, 0x2, 0x0, 0x5,
  0x0, 0x0, 0x0, 0x6f,
Response From Flood Sensor Node Received: Sensor Alarm Reported NO
  Flood Danger
```

We have added new cases to handle reports from security sensors for the Z-Wave message handler. In the next session, we will make an overview of the system.

Wrapping up the message parsing system

We have added new sensors to the Smart Home application. In order to handle all incoming messages correctly, we need to be able to parse messages correctly to inform the user. When we added new command classes, we needed to handle their values separately. A part of the message parser, which handles sensor commands, is shown here. We also used a part of this in the previous chapter to handle energy meter values such as illumination and temperature:

```c
int parse_incoming_mesage(uint8_t* message, int length) {
//Message Length
  uint8_t length_of_rest = message[1];
  uint8_t message_type = message[2];
  uint8_t message_function = message[3];

  else if (message_function == RESPONSE_RECEIVED) {
   printf("Response From ");
   if (message[5] == MultiSensorNodeID) {
     printf("Multi-Sensor Node Received: ");
   } else if (message[5] == WallPlugNodeID) {
     printf("Wall Plug Node Received: ");
   } else if (message[5] == LampHolderNodeID) {
     printf("Lamp Holder Node Received: ");
   } else if (message[5] == FloodSensorNodeID) {
     printf("Flood Sensor Node Received: ");
```

```
  }
  //Length of Data
  uint8_t data_length = message[6];
  uint8_t command_class = message[7];

  //Check Command Class and Take Action Accordingly
  if (command_class == COMMAND_CLASS_BATTERY) {
    int battery = 0;
    //If battery is not 0%
    if (message[9] != 0xFF) {
      battery = message[9];
    }
    if (message[5] == MultiSensorNodeID) {
      multi_sensor_battery_level = battery;
    } else if (message[5] == FloodSensorNodeID) {
      flood_sensor_battery_level = battery;
    }
    printf("Battery Level of Device is %d % \n", battery);
  } else if (command_class == COMMAND_CLASS_SENSOR_BINARY) {
    if (message[8] == BINARY_REPORT) {
      if (data_length == 0x04) {
        if (message[10] == DOOR_WINDOW_SENSOR) {
          if (message[9] == ON) {
            printf("Door/Window Sensor is OPEN\n");
            door_sensor_status = message[9];
          } else if (message[9] == OFF) {
            printf("Door/Window Sensor is CLOSE\n");
            door_sensor_status = message[9];
          }
        } else if (message[10] == MOTION_DETECTION_SENSOR) {
          if (message[9] == ON) {
            printf("Motion DETECTED\n");
            motion_sensor_status = message[9];
          } else if (message[9] == OFF) {
            printf("NO Motion Detected\n");
            motion_sensor_status = message[9];  }     }}

} else if (command_class == COMMAND_CLASS_SENSOR_MULTILEVEL_V2) {
    if (message[8] == MULTILEVEL_REPORT) {
       // Calculate Sensor Value
      int len = data_length - 3;
      uint8_t *data = malloc(len * sizeof(uint8_t));
      uint8_t scale;
      int i = 0;
```

```
      for (; i < len; i++) {
        data[i] = message[10 + i];
      }
      float value = calculate_sensor_value(data, len, &scale);
      if (message[9] == TEMPERATURE_SENSOR) {
        printf("Temperature Value is %f ", value);
        temperature = value;
        if (scale) {
          printf("Fahrenheit \n");
        } else {
          printf("Celcius \n");
        }
      } else if (message[9] == LUMINANCE_SENSOR) {
        printf("Illumination is %f ", value);
        illumination = value;
        if (scale) {
          printf("Lux \n");
        } else {
          printf("% \n");
        }
      } else if (message[9] == POWER_SENSOR) {
        printf("Power Value is %f \n", value);
        current_energy_consumption = value;
        if (scale) {
          printf("BUT/h \n");
        } else {
          printf("Watts \n");
        }
      }
    }
  } else if (command_class == COMMAND_CLASS_BASIC) {
    printf("Device is ");
    if (message[9] == OFF) {
      printf("Sleeping\n");
    } else if (message[9] == ON) {
      printf("Active\n");
    }
    if (message[5] == WallPlugNodeID) {
      wall_plug_status = message[9];
    } else if (message[5] == MultiSensorNodeID) {
      multi_sensor_status = message[9];
    } else if (message[5] == LampHolderNodeID) {
      lamp_holder_status = message[9];
    } else if (message[5] == FloodSensorNodeID) {
```

```
          flood_detector_status = message[9];
        }
      }
......//some code here............
.................. .
      } else if (command_class == COMMAND_CLASS_METER) {
        //Open ZWave Meter
        int len = data_length - 3;
        uint8_t *data = malloc(len * sizeof(uint8_t));
        uint8_t scale;
        int i = 0;
        for (; i < len; i++) {
          data[i] = message[10 + i];
        }
        float value = calculate_sensor_value(data, len, &scale);
        if (message[8] == METER_REPORT) {
          uint8_t meter_type = message[9] & 0x1F;
          if (meter_type == METER_ELECTRICITY) {
            const char* label = electricity_label_names[scale];
            const char* unit = electricity_unit_names[scale];
            printf("Electicity Meter Report %s %f %s\n", label,
            value, unit);
            if (scale == 2) {
              current_energy_consumption = value;
            } else if (scale == 0) {
              cumulative_energy_level = value;
            }
          }
        }
      } else if (command_class == COMMAND_CLASS_SENSOR_ALARM) {
        printf("Sensor Alarm ");
        if (message[8] == SENSOR_ALARM_REPORT) {
          printf("Reported ");
          if (message[10] == FLOOD_ALARM) {
            if (message[11] == ON) {
              printf("FLOOD Detected\n");
            } else if (message[11] == OFF) {
              printf("NO Flood Danger\n");
            }
            flood_sensor_status = message[11];
          } else if (message[10] == SMOKE_ALARM) {
            if (message[11] == ON) {
              printf("SMOKE Detected\n");
            } else if (message[11] == OFF) {
```

```
        printf("NO Fire Danger\n");
    }}}}}
  return 0;
}
```

We have also added a new command function to request an alarm sensor status. `get_sensor_alarm_value` makes the request to the corresponding alarm sensor to get its latest status:

```
int get_sensor_alarm_value(int serial_device, uint8_t nodeID,
uint8_t sensor_type, uint8_t callbackID) {
int message_length = 12;
uint8_t checksum = 0x00;
uint8_t message_buffer[] = { SOF, (message_length - 2), REQUEST,
SEND_DATA,
    nodeID, 0x03, COMMAND_CLASS_SENSOR_ALARM, SENSOR_ALARM_GET,
sensor_type,
    TRANSMIT_OPTION, callbackID, checksum };
checksum = generate_checksum(message_buffer, message_length);
message_buffer[message_length - 1] = checksum;
return write_to_serial_device(serial_device, message_buffer,
message_length);
}
```

As can be seen in the message parser function, when a sensor value or device status is received, it sets a static variable value, that will help us in the following chapter to monitor the current situation of the area. In order to monitor sensors and send them messages, a synchronous mechanism needs to be implemented carefully so as to not miss any messages from the controller.

In the next chapter, while we are in the process of improving our sample application, we will add a thread lock mechanism to use the Z-Wave USB adapter safely with multiple threads. That will help us monitor, send messages, and receive messages in a more secure way.

With the latest code samples, we will get experience at handling Z-Wave messages. This will help you to understand how complex it gets when the number of devices and cases increases, and so every case should be handled carefully, especially security cases; in real life scenarios, not doing so can cause damage in your home or any other residential areas where home automation has been setup.

Summary

During the course of this chapter, we have looked into sensors that can be used to develop a security device with Intel Galileo. These sensors are motion (PIR) sensors, door/window sensors, water sensors, and gas/smoke sensors. We have developed a simple application with the MQ-9 gas sensor to detect carbon monoxide in the air. In order to get the value of the carbon monoxide level in the air, we used analog pins in Intel Galileo.

Then we proceeded to cover some home security devices such as remote door/window sensors, motion sensors, smoke detectors, water leak detectors, and door lock sensors. We added a new sensor, a flood detector, to our Smart Home application system to detect floods in the residential area.

In the next chapter, we will investigate how we can use cameras with Intel Galileo for home surveillance. Then we will extend our application to control the camera in order to increase the security use case by coordinating other security sensors.

6
Home Surveillance and Extending Security Use Cases

In the previous chapter, we looked into devices and sensors used to secure your home. We will proceed by using another device at the heart of home automation, a camera. Cameras are the main devices used for security. They have been used in all kinds of places for security reasons. Government officials use security cameras to monitor streets and other public areas. Private companies monitor their offices. Cameras are also widely used in home automation.

The main reason to use security cameras for home automation is to detect if anyone broke into your house and capture their video or photo. They are also used to monitor a newborn baby by working parents. There are many providers who sell cameras with cloud services to monitor homes with a smartphone or any other connected device.

In this chapter, we will use a network camera to capture frames with Intel Galileo. Cameras will be the last device to be included in our home automation system. Following the use of cameras within the home automation system, we will improve the Smart Home application with new use cases and learn more about software development on Intel Galileo for home automation.

Introducing network cameras

Nowadays, there are plenty of manufacturers who sell network cameras that are able to connect to home network via Ethernet or Wi-Fi. Most of the companies also provide a free or paid cloud account for you to register your camera to their cloud system. After you've authorized your camera to stream to a private cloud server, you will be able to get a video stream from your home anywhere with an Internet connection.

Network cameras use the HTTP or RTSP Internet protocols to enable you to access a video stream from the camera. Most network cameras stream videos with the **MJPEG (Motion JPEG)** format over HTTP. MJPEG is a widely used format by digital cameras; it compresses sequences of JPEG images in a MJPEG container.

Most network cameras also have a web interface to access the camera via a local network and change settings. Within these settings, it also enables security with defined users wherein only users who have an account on the network cameras are allowed to access them and stream.

Throughout this chapter, we will use a network camera from D-Link DCS-930L, as shown in the following image, for our samples in the following section. We will try to obtain captures with Intel Galileo and see how to stream video coming from a network camera via Intel Galileo.

Some network cameras have a few built-in functions to be used for the security of your home. These built-in functions could be motion detection and sound detection. In accordance with the defined function, it can send a frame to the defined email client.

D-Link DCS-930L is a pure network camera without any other connection protocol defined. There are also network cameras powered with home automation protocols such as ZigBee and Z-Wave. These protocols are not used for video streaming; their purpose is to enable remote control on cameras. For example, you can rotate, zoom in, or zoom out a ZigBee camera remotely by sending a command via the ZigBee protocol.

There are plenty of network cameras with different features enabled. We will deal with the main feature of the camera to stream or capture photos.

Using cameras with Intel Galileo

Even though Intel Galileo is a headless device without any video output support or graphical processor unit, you can get video streams to Intel Galileo. In order to work with video cameras, either a USB webcam or a network camera, we will need the **OpenCV (Open Source Computer Vision)** library. If you work with an USB camera, you will also need the required Linux kernel modules (v4l2) to recognize video devices connected to Intel Galileo from the USB port.

Before getting into the details of OpenCV API and internals, we will build the OpenCV library and v4l2 module for Intel Galileo and make it ready for development with OpenCV. If you have not built OpenCV, you can follow the instructions given in the following section.

Building OpenCV and V4L2 for Intel Galileo

As Intel Galileo has a different processor, Intel Quark, from regular computers, we need to recompile and build any libraries with the cross-compiler and toolchain provided with Yocto Project. A successful cross-compile process will create necessary binaries for us to run on Intel Galileo.

In the *Chapter 1, Getting Started with Intel Galileo*, we initialized a build environment for Yocto Project in order to create a Linux filesystem and toolchain for Intel Galileo. Yocto Project allows you to build a full Linux filesystem or any other software without rebuilding the whole Linux filesystem. During the build environment initialization process, many Yocto Project recipes have been downloaded but not been used as they have not been included in the image. OpenCV recipe is one of the recipes downloaded during this process along with many others.

 If you've already made the full-image build in *Chapter 1, Getting Started with Intel Galileo,* it includes the OpenCV and video device drivers installed with the image.

You can locate the OpenCV Yocto Project recipe in $BUILD_DIRECTORY/meta-clanton_v1.0.1/meta-oe/meta-oe/recipes-support/opencv directory. opencv_2.4.9.bb is the file that includes instructions and rules to download the OpenCV source from an upstream repository and build. When you open the opencv_2.4.9.bb file, you will see that the line starts with DEPENDS; it is the variable that includes the OpenCV library, which needs to be built. v4l2-utils is the library that includes the kernel modules, so when we start building, OpenCV v4l2 modules will be built together.

```
$ cd $BUILD_DIRECTORY/meta-clanton_v1.0.1
$ cd source poky/oe-init-build-env yocto_build
$ bitbake opencv
```

> The previous build process should work to build OpenCV for the Intel Galileo Quark processor without any problem. In some cases, the OpenCV recipe might not be updated and you could have some build errors. In the event of an error, you can check the upstream source code link; MD5 checksum in the recipe.

OpenCV ipk packages are listed under $BUILD_DIRECTORY/meta-clanton_v1.0.1/yocto_build/tmp/deploy/ipk/i586. The v4l2 kernel module packages are in the $BUILD_DIRECTORY/meta-clanton_v1.0.1/yocto_build/tmp/deploy/ipk/clanton directory.

In order to install related packages, you can copy OpenCV, the v4l2 packages, and their dependency module packages to the Intel Galileo filesystem and then install them with the help of opkg tool. In addition to installing the packages to Intel Galileo, you also need to install development files to the Intel Galileo toolchain directory on your host to compile and build your OpenCV application. In *Chapter 1, Getting Started with Intel Galileo*, we installed SDK to the /opt/clanton-tiny/1.4.2/ directory. You need to install development files into the /opt/clanton-tiny/1.4.2/sysroots/i586-poky-linux-uclibc directory.

However, instead of using a copy and install process, you can add OpenCV to the image recipe and that will automatically install the required packages to the filesystem image.

Edit $BUILD_DIRECTORY/meta-clanton_v1.0.1/meta-clanton-distro/recipes-core/images/image-full.bb, add a new line as IMAGE_INSTALL += "opencv", and rebuild the image and toolchain. Then follow the instructions from *Chapter 1, Getting Started with Intel Galileo*, to copy the required files to the SD card and re-install the toolchain into your host machine supporting OpenCV.

You can add any application you want to your custom Linux image built by Yocto Project with the defined method.

Introducing OpenCV

In order to work with images or video coming from your security camera, you need to know the basics of the OpenCV library. OpenCV is an open source cross-platform computer vision and machine learning library. OpenCV provides hundreds of algorithms for analytics on images, mostly for real-time use cases, and that's why it is used in robotics appliances.

 This link is the OpenCV Official website: http://opencv.org. This link has the OpenCV Offical documentation: http://docs.opencv.org.

Visual analytics algorithms require a vast amount of processing power. Therefore, OpenCV requires a powerful CPU and GPU for complex analytics. However, we will only use some basic functionality from OpenCV just for surveillance purposes. Intel Galileo's CPU is powerful enough to do basic image manipulation such as saving a captured image or video from a camera. It will be enough for home surveillance. If you would like to do more complex analytics with your home automation application such as face recognition, Intel Galileo may lack the required computational power.

 OpenCV API is essentially a C++ API, but for basic functionality you can still use C API. It is suggested you use C++ API for this purpose. Yocto Linux also has a C++ compiler and libraries, and so you can also use the OpenCV C++ API with it.

Let's briefly overview the functionality we will use from OpenCV to capture frames from a network camera or USB camera.

highgui – high-level GUI and media I/O

The **highgui** module provides functions for developers to create **graphical user interface (GUI)** elements for applications. This helps developers to create windows to show the captured video or a frame. Another useful functionality of this module is to provide an interface for developers to read video from a USB camera, a network camera, or a video file on disk. While helping developers to read video, it also provides functions to write or save a picture or a video to a file on storage devices.

Since Intel Galileo doesn't have video output, we do not need to use GUI functionality. However, we definitely need to be able to read from video devices for home surveillance.

We will use the following functions from the OpenCV highgui module:

- C++: This is the `VideoCapture` module that captures video from a USB device or network camera.

  ```
  VideoCapture cap(0) //Capture from /dev/video0
  VideoCapture cap("http://network.cam.ip/stream.mjpeg")
    //Capture from network camera
  ```

- C: In the C API, we need to call the `cvCaptureFromCAM(int index)` function to get streams from a USB camera; we need to use `cvCaptureFromFile(const char* filename)` to capture streams from a network camera.

  ```
  cvCaptureFromCAM(0) //Capture from /dev/video0
  cvCaptureFromFile("http://network.cam.ip/stream.mjpeg")
  ```

- C++: The `imwrite(const string& filename, InputArray img, const vector<int>& params=vector<int>())` function can be used to save an image to disk in a C++ application.

  ```
  imwrite("capture.jpeg",image);
  ```

- C: The `cvSaveImage(const char* filename, const CvArr* image, const int* params=0 )` function can be used to save an image to disk in a C application:

  ```
  cvSaveImage("capture.jpeg",image,0);
  ```

- C++: The `VideoWriter(const string& filename, int fourcc, double fps, Size frameSize, bool isColor=true)` function can be used to save video stream to disk in a C++ application. A basic use of this is shown here:

  ```
  Size S = Size( (int)capture.get(CV_CAP_PROP_FRAME_WIDTH),
              (int)capture.get(CV_CAP_PROP_FRAME_HEIGHT));
  VideoWriter  vWriter("video_output.avi",
  CV_FOURCC('P','I','M','1'), 20, S, true);
  vWriter(frame);
  ```

- C: The `cvCreateVideoWriter(const char* filename, int fourcc, double fps, CvSize frame_size, int is_color=1 )` function can be used to save video streams to disk in a C application.

  ```
  CvSize S;
  s.width = iwidth;
  s.height = height;
  ```

```
CvVideoWriter *writer =
cvCreateVideoWriter("video_output.avi",
    CV_FOURCC('M', 'J', 'P', 'G'),  20, S);
```

We will mostly use the basic functionalities mentioned previously from OpenCV. Along with the functions mentioned previously, we will also need to use data structures that store frames and their properties. It is suggested you use the C++ API but, to bind with all the samples we have developed during the book, we will not develop a complicated application, and so we will do basic C applications in the later sections.

Let's follow up with sample applications to capture and store images from Intel Galileo.

Capturing images from a camera with Intel Galileo

We will have two sample applications to capture images from a USB camera and the D-Link DSC-930L network camera and save them to Intel Galileo.

We started with creating a new C Project, as described in *Chapter 1*, *Getting Started with Intel Galileo*, when we created a C file as capture_from_usb.c. We need to add OpenCV libraries to Makefile in order to build application with OpenCV libraries. Makefile.am should look like this:

```
bin_PROGRAMS = capture_from_usb
capture_from_usb_SOURCES = capture_from_usb.c
AM_CFLAGS = @capture_from_usb_CFLAGS@
AM_LDFLAGS = @capture_from_usb_LIBS@ -lopencv_core -lopencv_imgproc -
    lopencv_legacy -lopencv_ml -lopencv_highgui -lm
CLEANFILES = *~
```

While capturing from USB devices, we need to provide the index of the video file. In our case, the video file is /dev/video0, and so we put the index as 0.

```
#include <stdlib.h>
#include <stdio.h>
#include "opencv/cxcore.h"
#include "opencv/cv.h"
#include "opencv/highgui.h"
#define USBCAM 0
int main(void) {
        //Get USB Cam Device
```

```
        CvCapture *pCapturedImage = cvCreateCameraCapture(USBCAM);
        //Get Frame from Capture Device
        IplImage *pSaveImg = cvQueryFrame(pCapturedImage);
        //Save Image to Filesystem
        cvSaveImage("capture.jpg", pSaveImg, 0);
        //Release Image Pointer
        cvReleaseImage(&pSaveImg);
        return 0;
    }
```

When we run the application `capture_from_usb` binary inside the Intel Galileo, it will create the `capture.jpeg` file.

root@clanton:~# ./capture_from_usb

Let's create a new project to capture from a network camera. In the new project, we created `capture_from_network.c` and build with Intel Galileo toolchain.

While accessing streams from a network camera, we need to provide the IP address of the network camera. In our case, it is 192.168.2.141 in our local network. In most cases, network cameras are secured with user credentials. We also provided the username and password as admin and 123 respectively with the link with `IP_CAM`, as seen in the sample code given here:

```
#include <stdlib.h>
#include <stdio.h>
#include "opencv/cxcore.h"
#include "opencv/cv.h"
#include "opencv/highgui.h"
#define IP_CAM "http://admin:123@192.168.2.141:80/mjpeg.
cgi?user=admin&password=1 23&channel=0&.mjpg"
int main(void) {
  //Get Capture Device assign to a memory location
  CvCapture *pCapturedImage = cvCaptureFromFile(IP_CAM);
  //Frame to Save
  IplImage *pSaveImg = cvQueryFrame(pCapturedImage);
  //Save Image to Filesystem
  cvSaveImage("capture.jpg", pSaveImg, 0);
  //Release Memory Pointer
  cvReleaseImage(&pSaveImg);
  return 0;
}
```

When we run the application `capture_from_network` inside Intel Galileo, it will create the `capture.jpeg` file:

root@clanton:~# ./capture_from_network

You can copy the `.jpeg` file to your host machine to open and view the frame. Just like copying files to Intel Galileo from your host machine using the SCP tool, you can use SCP to transfer files to your host machine using your host machine's credentials.

Saving a video from a camera with Intel Galileo

Let's follow up from the previous section with saving video, instead of only one frame, to Intel Galileo. We will get the video from a network camera and will save it to Intel Galileo.

We have created a new project and a new `.c` file `save_video.c` in the project. We will use the same libraries that were used in capturing an image. Our aim is to record 5 seconds of video to Intel Galileo, as shown here:

```
#include <stdlib.h>
#include <stdio.h>
#include "opencv/cxcore.h"
#include "opencv/cv.h"
#include "opencv/highgui.h"
#include <time.h>
#define IP_CAM
  "http://admin:123@192.168.2.141:80/mjpeg.cgi?user=admin&password=1
  23&channel=0&.mjpg"
#define FPS 15
#define DURATION 5 //seconds
int main(void) {
  //Capture Device and Captured Frame
  CvCapture *pCapture = cvCaptureFromFile(IP_CAM);
  int width = (int) cvGetCaptureProperty(pCapture,
  CV_CAP_PROP_FRAME_WIDTH);
  int height = (int) cvGetCaptureProperty(pCapture,
  CV_CAP_PROP_FRAME_HEIGHT);
  //Get Size of Video
  CvSize S;
  S.height = height;
  S.width = width;
  //Frame Structure
  IplImage *frame;
  //Create VideoWriter
```

```
CvVideoWriter* videoWriter = cvCreateVideoWriter("record.avi",
CV_FOURCC('X', 'V', 'I', 'D'), FPS, S, 1);
//Get Current Time
time_t now;
struct tm *tm;
now = time(0);
//Start Recording for Given Time
printf("Recording....\n");
if ((tm = localtime(&now)) == NULL) {
  printf("Can't Get Time\n");
  return -1;
}
int start = tm->tm_sec;
int progress = tm->tm_sec;
while ((progress - start) <= DURATION) {
  frame = cvQueryFrame(pCapture);
  cvWriteFrame(videoWriter, frame);
  now = time(0);
  if ((tm = localtime(&now)) == NULL) {
    printf("Can't Get Time\n");
    return -1;
  }
  progress = tm->tm_sec;
}
printf("Saved....\n");
//Release Pointers from Memory
cvReleaseCapture(&pCapture);
//This is needed to make recorded video container to be closed
properly
cvReleaseVideoWriter(&videoWriter);
return 0;
}
```

When we have run the `save_video` binary, it will create a `record.avi` file. You can copy the file to your host machine and view the recorded video with a video player:

```
root@clanton:~# ./save_video
Recording ....
Saved ....
```

Saving video to Intel Galileo is not recommended as it has limited space. The initial state of the filesystem on the SD card will not allow you to save larger files or use all the space on the SD card. It is advised to resize the filesystem image on the SD card. First mount your SD card on your Linux host machine as shown in the following command:

```
$ sudo mount /dev/sdc1 /media/sdcard
```

Run a filesystem check on the current filesystem image file as shown in the following command:

```
$ sudo fsck.ext3 -f /media/sdcard/image-full-galileo-
  clanton.ext3
```

Finally, you can resize with the following command. In our example, we resize our image to 4-gigabyte, by supplying 4-gigabyte as kilobytes:

```
$ sudo resize2fs /media/sdcard/image-full-galileo-
clanton.ext3 1638400
```

Now you can unmount the SD card from the host machine and can boot on Intel Galileo.

Streaming a video from Intel Galileo

Accessing the video camera in your home remotely is essential for home security. If you want to connect your Intel Galileo to a cloud system and access a video stream through your Intel Galileo, you need to stream video from Intel Galileo. You may also want to stream video from a USB device and access it remotely; for this you would need to create an application, or a worker thread in your application, to stream the video.

Since Intel Galileo doesn't have GPU or any other supported video encoders/ decoders, it will be wise not to try any transcoding to change the video format. A suggested method is to stream video from Intel Galileo, writing the JPEG frames to a network socket, and have another device read the frames.

You can check the Linux socket programming for more details.

Adding new use cases

We have finished working with devices for home automation. We covered almost all the common devices used for home automation. As we have completed adding a camera, we have successfully achieved the state where we can monitor a home.

Now we will proceed to make a few last additions to the Smart Home application such as adding some rules; this includes completing some actions when a certain device value is changed,. For example, when motion is detected, we will make the Smart Home application perform a capture.

Let's first include the camera module in the Smart Home application. We have added the `camera.c` and `camera.h` files to manage a network camera and also changed the `Makefile.am` file to link OpenCV libraries for the build process. The following header file can be used within the application:

```
#ifndef CAMERA_H_
#define CAMERA_H_
#define IP_CAM "http://admin:123@192.168.2.141:80/mjpeg.
cgi?user=admin&password=1
  23&channel=0&.mjpg"
#define USBCAM 0
/**
 * Capture Frame from Network Camera if defined else from USB
   Camera
 * @param void
 * @return void
 */
void capture_frame(void);
/**
 * Record Video from Network Camera if defined else from USB
   Camera
 * @param int duration in seconds
 * @return void
 */
void record_video(int sec);

#endif /* CAMERA_H_ */
```

We made a small change in our code as previously shown in the sample camera applications. We used a predecessor to check whether we have defined the IP camera video URL for capture, for conditional compilation of the code. If we don't define it, the compiler will automatically build the application for a USB camera. After we add the predecessor, the `capture_frame` function looks like this:

```
void capture_frame() {
  //Get Capture Device assign to a memory location
#ifdef IP_CAM
  CvCapture *pCapturedImage = cvCaptureFromFile(IP_CAM);
#else
  CvCapture *pCapturedImage = cvCreateCameraCapture(USBCAM);
```

```
#endif
  //Frame to Save
  IplImage *pSaveImg = cvQueryFrame(pCapturedImage);
  //Get TimeStamp
  //Get Current Time
  time_t now;
  struct tm *tm;
  now = time(0);
  if ((tm = localtime(&now)) == NULL) {
    printf("Can't Get Time\n");
    return;
  }
  char buf[256];
  printf("Capturing....\n");
  sprintf(buf, "capture%d_%d_%d_%d_%d_%d.jpg", tm->tm_year, tm-
>tm_mon,
      tm->tm_mday, tm->tm_hour, tm->tm_min, tm->tm_sec);
  //Save Image to Filesystem
  cvSaveImage(buf, pSaveImg, 0);
  printf("Saved....\n");
  return;
}
```

As we set the device control in the `capture_frame` function, we also implemented the same setting in the `record_video` function.

Let's proceed to add some conditions to our application; let it capture frames if motion is detected or the door sensor is opened.

Adding new rules

While developing the Smart Home application, we added all the devices connected to Intel Galileo. In order to handle periodic updates from devices, we created a thread to query the devices, get the latest updates from them and, according to the updated value, take the necessary emergency action. We called this module device and added `device.c` and `device.h` files. The `device.h` file includes the defined macros, stores the latest status of the devices, and records it to a XML file for external access to device statuses.

```
#ifndef DEVICE_H_
#define DEVICE_H_
#include "message_queue.h"
/**
 * Transmit Option for ZWave Controller
 */
```

```
#define TRANSMIT_OPTION       0x25  //Aeon USB Stick
/**
 * Staus Variable Buffers
 */
#define BUFFER_MAX          256
#define FILE_LINE         1024
/**
 * Define Update Frequency
 */
#define UPDATE_FREQUENCY    2   //Minutes
/**
 * Device Status XML Filename
 */
#define XML_FILE_NAME       "/home/root/smarthome/home.xml"
#define JSON_FILE_NAME      "/home/root/smarthome/home.json"
/**
 * Constant Strings
 */
#define status_sleeping     "Sleeping"
#define status_active       "Active"
#define device_on          "ON"
#define device_off         "OFF"
#define detected           "DETECTED"
#define not_detected       "NOT DETECTED"
#define door_open          "OPEN"
#define door_closed        "CLOSED"
/**
 * Defined Nodes
 */
#define NumberofNodes       5
#define ControllerNodeID    0x01
#define MultiSensorNodeID   0x02
#define WallPlugNodeID      0x03
#define LampHolderNodeID    0x04
#define FloodSensorNodeID   0x05
/**
 * Device Names
 */
#define TemperatureSensorName     "SHT11 Sensor"
#define GasSensorName           "MQ-9 CO Sensor"
#define ControllerNodeName       "AeonUSB Stick"
#define MultiSensorNodeName      "Philio Multi-Sensor"
#define WallPlugNodeName        "Fibaro Wall Plug"
#define LampHolderNodeName       "Everspring Lamp Holder"
#define FloodSensorNodeName      "Everspring Flood Detector"
```

```
#define NetworkCameraName        "D-Link Network Camera"
/**
 * Device Status for ZWave Nodes
 */
char multi_sensor_status[BUFFER_MAX];
char wall_plug_status[BUFFER_MAX];
char lamp_holder_status[BUFFER_MAX];
char flood_detector_status[BUFFER_MAX];
char temperature_sensor_status[BUFFER_MAX];
char gas_sensor_status[BUFFER_MAX];
/**
 * Power Level Statuses for ZWave Nodes
 */
char multi_sensor_power_level[BUFFER_MAX];
char wall_plug_power_level[BUFFER_MAX];
char lamp_holder_power_level[BUFFER_MAX];
char flood_detector_power_level[BUFFER_MAX];
/**
 * Battery Levels
 */
int multi_sensor_battery_level;
int flood_detector_battery_level;
/**
 * Current Environment Status
 */
float temperature_f;
float temperature_c;
float relative_humidity;
float illumination;
float co_level;
/**
 * Energy Meter
 */
float current_energy_consumption;
float cumulative_energy_level;
/**
 * Switch Status
 */
char lamp_holder_switch[BUFFER_MAX];
char wall_plug_switch[BUFFER_MAX];
/**
 * Security Sensor Status
 */
```

```
    char flood_sensor_status[BUFFER_MAX];
    char door_sensor_status[BUFFER_MAX];
    char motion_sensor_status[BUFFER_MAX];
    /**
     * Camera Related Status
     */
    char network_camera_status[BUFFER_MAX];
    char network_camera_port[BUFFER_MAX];
    void* update_status(void* arg);
    void print_device_xml(const char* file_name);
    #endif /* DEVICE_H_ */
```

The `update_status` thread is our worker thread that requests updates from Z-Wave devices, reads from the MQ-9 Gas Sensor, the SHT11 temperature and humidity sensor, and finally takes emergency actions.

```
    /**
     * Periodic update worker
     * @param null
     * @return null
     */
    void* update_status(void* arg) {
      set_defaults();
      while (1) {
        /**
         * Periodic Updates from Connected Devices
         */
        /**
         * Philio Multi-Sensor Node Commands
         */
        get_device_status( MultiSensorNodeID, 1);
        get_battery_level( MultiSensorNodeID, 2);
        get_binary_sensor_value( MultiSensorNodeID,
      DOOR_WINDOW_SENSOR, 3);
        get_binary_sensor_value( MultiSensorNodeID,
      MOTION_DETECTION_SENSOR, 4);
        get_multilevel_sensor_value( MultiSensorNodeID,
      TEMPERATURE_SENSOR, 5);
        get_multilevel_sensor_value( MultiSensorNodeID,
      LUMINANCE_SENSOR, 6);
        get_node_power_level( MultiSensorNodeID, 7);
        /**
         * Fibaro Wall Plug Commands
         */
        get_device_status( WallPlugNodeID, 8);
```

```
get_binary_switch_status( WallPlugNodeID, 9);
get_meter_level( WallPlugNodeID, POWER, 10);
get_meter_level( WallPlugNodeID, ENERGY, 11);
get_node_power_level(WallPlugNodeID, 12);
/**
 * Everspring Lamp Holder Commands
 */
get_device_status( LampHolderNodeID, 13);
get_binary_switch_status( LampHolderNodeID, 14);
get_node_power_level(LampHolderNodeID, 15);
/**
 * Everspring Flood Detector Commands
 */
get_device_status( FloodSensorNodeID, 16);
get_sensor_alarm_value( FloodSensorNodeID, FLOOD_ALARM, 17);
get_node_power_level(FloodSensorNodeID, 18);
/**
 * Read From Sensors Connected to Intel Galileo
 * SHT11 and MQ-9
 */
delaySeconds(2);
snprintf(temperature_sensor_status, sizeof(status_active),

status_active);
temperature_c = read_temperature();
    delaySeconds(2);
relative_humidity = read_humidity();
delaySeconds(2);
snprintf(gas_sensor_status, sizeof(status_active),
status_active);
co_level = read_gas_sensor();
/**
 * Check for Emergency Actions
 */
emergency_actions();
/**
 * Sleep for 2 Minutes Print Report to devices.xml file
 */
update_device_xml(XML_FILE_NAME);
update_device_json(JSON_FILE_NAME);
delayMinutes(UPDATE_FREQUENCY);

}
return NULL;
}
```

The `emergency_actions` function is an enclosed function inside the `device.c` file to complete emergency actions as given in the following code snippet:

```
void emergency_actions() {
  //If There is a Flood Switch Off Wall Plug
  if (strcmp(flood_detector_status, detected) == 0) {
    if (strcmp(wall_plug_switch, device_on) == 0) {
      binary_switch_on_off( WallPlugNodeID, OFF, 0x11);
    }
  }
  //If There is a Gas Leank Switch Off Wall Plug
  if (gas_voltage > GAS_THRESHOLD) {
    if (strcmp(wall_plug_switch, device_on) == 0) {
      binary_switch_on_off( WallPlugNodeID, OFF, 0x11);
    }
  }
  //If There is a motion capture from Network Camera
  if (strcmp(motion_sensor_status, detected)) {
    capture_frame();
  }
  //If Door Window Sensor is Open
  if (strcmp(door_sensor_status, door_open)) {
    capture_frame();
  }
}
```

As we've seen in the `emergency_actions` function, when there is motion detection or the door/window sensor is open, the Smart Home application will capture an image from the network camera. This is a very basic security precaution to capture images where there is motion, as this isn't supposed to happen.

We may also add the `capture_frame` function commands to the Z-Wave message handler as we did in the previous section to directly capture a frame when motion is detected. First, we need to include the `camera.h` file into the `message.c` file for this. Then, we can add the `capture_frame` function, as shown in the following code:

```
//Code section from message.c
if (message[10] == DOOR_WINDOW_SENSOR) {
        if (message[9] == ON) {
            printf("Door/Window Sensor is OPEN\n");
            snprintf(door_sensor_status, sizeof(door_open),
                door_open);
            capture_frame();
        } else if (message[9] == OFF) {
```

```
        printf("Door/Window Sensor is CLOSE\n");
        snprintf(door_sensor_status, sizeof(door_closed),
            door_closed);
    }
} else if (message[10] == MOTION_DETECTION_SENSOR) {
    if (message[9] == ON) {
        printf("Motion DETECTED\n");
        snprintf(motion_sensor_status, sizeof(detected),
            detected);
        capture_frame();
    } else if (message[9] == OFF) {
        printf("NO Motion Detected\n");
        snprintf(motion_sensor_status, sizeof(not_detected),
            not_detected);
    }
}
}
.............. //code section end from message.c
```

Summary

In this chapter, you learned how to work with a camera with Intel Galileo. Before developing applications with Intel Galileo, we have provided an introduction to the OpenCV library. We have briefly looked at the OpenCV library and at how we can build the library for Intel Galileo Yocto Linux.

Then we developed simple applications to capture images and record video from USB and network cameras. After having enough practice with cameras, we added a new module to the Smart Home application to add some new security features such as getting captures when we detect a motion or when a door sensor is open.

In the following chapter, we will make our Smart Home application work as a service application or a background process. We will add new features to the Smart Home application and use the Yocto Project to define the service recipe file in order to add an application as a service.

7
Building Applications and Customizing Linux for Home Automation

In the previous chapters, you learned how to develop applications that work on Intel Galileo. You learned how to deploy applications to Intel Galileo with the help of Eclipse IDE or by using the SCP command-line tool to transfer application binary files. During the course of this chapter, we will work on how we can make our application or applications start at boot time and run as a background process or service on Intel Galileo Yocto Linux operating system.

While we arrived around the end of the development process, we would like to run our application in Intel Galileo when the board boots up with no user command to start it. We want the Smart Home application to run automatically because managing a home from just a command line terminal interface is not so user friendly. Therefore, for this, we will also need to define some features for the Smart Home application to send commands to and receive data from.

Let's start with what we can do to make Smart Home application start as a background process in Intel Galileo.

Customizing Linux with the Yocto Project

Customizing Linux means changing the default system configurations and adding or removing applications and kernel modules to the default Linux build for Intel Galileo. While following the steps described in the *Chapter 1, Getting Started with Intel Galileo*, we didn't make any changes, and so we only built the default full image defined for Intel Galileo.

In the previous chapters, we briefly mentioned adding the OpenCV library and added the cp210x kernel module to our default Linux image. The OpenCV library is an open source package and it was already delivered with the Intel Galileo board support package. In this section, we will make the required configurations and changes to build a structure to add our own application to the Linux image.

Adding a new application to Yocto Project

To add a new application, we need to add a Yocto Project recipe to the Intel Galileo Yocto Project. If you have developed a whole software stack with a number of applications, services, and libraries, it would be better to create your own directory like meta-clanton-distro and add the required configuration files to the directory. Finally, we need to add the new directory in the setup.sh file to the BBLAYERS variable. Configuration files make the BitBake script read the whole directory under the defined directory, parse the recipes, and build all of them. We will only deploy one application to the Intel Galileo default Yocto Linux image. Therefore, we will work only on the existing recipes and create our own Yocto Project recipe. Before going forward, it will be a good practice to review *Chapter 1*, *Getting Started with Intel Galileo*, to remember the steps you have initialized for the Yocto Project build environment. The build directory (BUILD_DIR) was /home/onur/galileo_build, and the Yocto Project files were stored in meta-clanton_v1.0.1.

Let's go into the build directory to create a new file and start writing the recipe, as shown in the following commands:

```
$ cd $BUILD_DIR/meta-clanton_v1.0.1
```

1. First, we need to create a directory for the recipe in /meta-oe/meta-oe/recipes-support. Directories such as recipes-multimedia, recipes-core, recipes-extended, and others, can also be selected. How you classify your application depends on you. Now create a directory using the following command:

   ```
   $ mkdir meta-oe/meta-oe/recipes-support/smarthome
   ```

2. Create the .bb file with which we will define our application for the build process; save it empty with *Ctrl + O*:

   ```
   $ nano meta-oe/meta-oe/recipes-support/smarthome_0.0.1.bb
   ```

3. We will start by entering the description for the application, application section, and a library or a base. Then you need to enter the software license type; like all the code in this book, this application is also licensed under the MIT license, license file checksum, and build dependency.

Generate the checksum for the COPYING file under the project folder. You can generate the checksum with the md5sum tool as shown in the following command:

```
$ md5sum COPYING
```

We defined the first lines of smarthome_0.0.1.bb file which are given here:

```
DESCRIPTION = "Home Automation Application for Intel Galileo"
SECTION = "base"
LICENSE = "MIT"
LIC_FILES_CHKSUM = "file://COPYING;md5=c5bb609535f48d5cd02fe14a780
f3d8c"
DEPENDS += "opencv"
```

4. Now, we will proceed to define the source code path to fetch the source code. In Yocto Project, recipes define the upstream source path. BitBake script downloads the source code from the given path and builds the downloaded source code. Paths are defined to the SRC_URI variable. Paths can be a Git, SVN repository, HTTP, FTP, or a local path. In the sample recipes, we defined the source in a local path. We also need to enter the checksum and sha256sum values. We added the following lines to our recipe:

```
SRC_URI = "file:///home/onur/tmp/galileo_ha/smarthome-0.0.1.tar.
gz"
SRC_URI[md5sum] = "46ef1371208ee57f89cbfa793a689eba"
SRC_URI[sha256sum] =
"dd4d8cd85c86c440173e60af36e227fbc2cffb151877ebaed8d474625cd889cd"
```

 You can use sha256sum Linux to generate a sha256sum value with the following commands on your host machine:

```
$ sha256 /home/onur/tmp/galileo_ha/smarthome-
0.0.1.tar.gz
```

5. In this step, we will define how to build our application. First, we define the source directory where the source code will be extracted. Then, we define how the BitBake script will build the application. We defined in the recipe to use autotools for the build process. We created the applications in Eclipse by selecting autotools. Autotools configurations are already defined in Yocto Project, and so we inherited configurations from an existing project to the recipe. See the next two lines for descriptions:

```
S = "${WORKDIR}/${PN}-${PV}"
inherit autotools gettext
```

6. Finally, we define the installation steps to make our home automation application automatically start and write outputs to a defined file. The BitBake script will use the defined autotools installation process and it will install the application into /usr/bin. The following definition will carry out extra installation steps for auto start descriptions. We will add the following lines at the end of the smarthome_0.0.1.bb file:

```
do_install_append(){
  mkdir ${D}/etc
  mkdir ${D}/etc/init.d
  mkdir ${D}/etc/rcS.d
  install -m 0755 ${S}/scripts/startha.sh ${D}${sysconfdir}/init.d
  ln -sf ../init.d/startha.sh ${D}${sysconfdir}/rcS.d/S99startha.sh
}
```

As you see in the previous installation definition, we also have a startup script to execute at boot time. We have included the script to the project source folder under scripts folder as startha.sh. The startup script is defined as given in the following command:

```
#!/bin/sh
# Start the Home Automation Application
echo "Starting Home Automation Application" > /var/log/homelog
/usr/bin/smart_home &> /var/log/homelog
```

7. The final content of the Yocto Project recipe for Smart Home application is as follows:

```
DESCRIPTION = "Home Automation Application for Intel Galileo"
SECTION = "base"
LICENSE = "MIT"
LIC_FILES_CHKSUM = "file://COPYING;md5=c5bb609535f48d5cd02fe14a780f3d8c"
DEPENDS += "opencv"
SRC_URI = "file:///home/onur/tmp/galileo_ha/smarthome-0.0.1.tar.gz"
SRC_URI[md5sum] = "46ef1371208ee57f89cbfa793a689eba"
SRC_URI[sha256sum] =
"dd4d8cd85c86c440173e60af36e227fbc2cffb151877ebaed8d474625cd889cd"
S = "${WORKDIR}/${PN}-${PV}"
inherit autotools gettext
do_install_append(){
  mkdir ${D}/etc
  mkdir ${D}/etc/init.d
  mkdir ${D}/etc/rcS.d
```

```
  install -m 0755 ${S}/scripts/startha.sh ${D}${sysconfdir}/init.d
  ln -sf ../init.d/startha.sh ${D}${sysconfdir}/rcS.d/S99startha.
sh
}
```

8. We are done with creating our Yocto Project recipe. It is time to build our new image and start it on Intel Galileo. Run the following commands inside the `meta-clanton_v1.0.1` folder to build Smart Home application and create the new image:

```
$ source poky/oe-init-build-env yocto_build
$ bitbake image-full-galileo
```

 You can follow the defined steps in *Chapter 1, Getting Started with Intel Galileo*, to copy the created files to an SD card and boot the image on Intel Galileo.

 You can apply the preceding steps to any application you have to add into the custom Linux image. To gain more expertise, you can read the *Yocto Project Developer's Manual* at `http://www.yoctoproject.org/docs/current/dev-manual/dev-manual.html`.

Adding new features to the application

We have gone through the steps to add an application to the Linux image running on Intel Galileo. If you have a service process running in the background, you should work out the ways in which you communicate with the process, to send commands to the application.

There are many methodologies that you can apply or implement for your application to receive commands and send outputs to the requested client. We will investigate two methodologies to send and receive messages to and from the application. We will use named pipes and network sockets. Named pipes help you deliver messages locally to the home automation service. Network sockets give you the ability to deliver messages through the network connection.

Using named pipes

A named pipe is usually used to create an inter-process communication structure between processes. The idea is to create a special Linux device file to read and write data and bytes to the named file. Named pipes use the **first in first out** (**FIFO**) principle. The first message in the FIFO will be received by the process in the named pipe.

We will follow the following steps to create a named pipe in the filesystem and read/write messages from the pipe.

1. We have to create a `pipe` file. If you create a pipe from the command line, you can use the `mknod` tool and set its permissions for the `chmod` tool. See the example shown here:

   ```
   root@clanton:~# mknod test p
   root@clanton:~# chmod 666 test
   root@clanton:~# ls -all test
   prw-rw-rw-    1 root      root              0 Jan  2 05:50
     test
   ```

   If you want to create the pipe in the application, here is the sample code:

   ```
   /* Create the FIFO if it does not exist */
   umask(0);
   mknod("PIPE_FILE", S_IFIFO|0666, 0);
   ```

2. We created our named pipe, so we need to be able to read the incoming messages from the pipe. We will use standard file libraries to open and read from the named pipe. A basic reading can be done with the following code:

   ```
   FILE *fp;
   char readbuf[80];
   fp = fopen(FIFO_FILE, "r");
   fgets(readbuf, 80, fp);
   printf("Received string: %s\n", readbuf);
   fclose(fp);
   ```

3. In order to write to the pipeline, you can simply use the `echo` tool or you can use the standard file library to write into the file you've opened.

When you execute the following command from the command line interface, it will send the show home status message to the SMARTHOMEPIPE variable and the application, which has opened the pipe, will receive it.

```
root@clanton:~# echo "show home status" > SMARTHOMEPIPE
```

You can write in a named pipe in a C application as shown in the following code:

```
FILE *fp;
if((fp = fopen(FIFO_FILE, "w")) == NULL) {
    perror("fopen");
    exit(1);
}
fputs(argv[1], fp);
fclose(fp);
```

Using named pipes in the application

In the application, we have created a thread to receive messages from the named pipe. In our application, we only needed to receive commands as we will write the output to an XML file. The thread's implementation is shown in the next code:

```
/**
 * Handle User Requests from PIPE
 */
void* named_pipe_handler(void* arg) {
  FILE *fp;
  char readbuf[USERBUF];

  /* Create the FIFO if it does not exist */
  umask(0);
  mknod(FIFO_FILE, S_IFIFO | 0666, 0);
  fp = fopen(FIFO_FILE, "r");
  while (1) {
    if (fgets(readbuf, USERBUF, fp) != NULL) {
      snprintf(command, sizeof(readbuf), readbuf);
    }
  }
  fclose(fp);
  return NULL;
}
```

Network sockets

Network sockets are extremely common in today's connected world to create communication between two devices connected to the Internet. Sockets are the virtual endpoints of a system on a network. You can imagine them as pipelines used from two different processes on different devices. If you want to communicate, Intel Galileo and an Android phone can send and receive messages through allocated network sockets, which is like using pipelines in the system.

You will use network sockets to enable your home automation application to receive and send messages from/to remote devices. This ability will enable your application to receive from any device with an Internet connection. For example, if you want to manage your Smart Home application with an Android device, you need to implement the code for the Android application to connect the defined network socket and send defined commands to our application such as switching a wall plug or lamp holder on and off.

We have defined a new module in the application named socket_listener. We have added the socket_listener.c and socket_listener.h variables to define a new method to work as a new thread to listen the defined network socket and receive commands from the outside.

```
#ifndef SOCKET_LISTENER_H_
#define SOCKET_LISTENER_H_

#define SOCKET 3500
#define MAXBUF 5000

void* socket_worker(void* arg);

#endif /* SOCKET_LISTENER_H_ */
```

The socket_worker function takes one argument, which is the command variable from the main function. The thread writes the received bytes to the command and the main function checks the received message. The content of the socket_listener.c file that includes the code for the socket_worker function is given here; the code for the socket_worker function is listening to the defined socket, which is 3000 in this case, and sends ACK as an acknowledgment.

```
#include "socket_listener.h"
#include <stdio.h>
#include <stdlib.h>
#include <string.h>
#include <unistd.h>
#include <arpa/inet.h>
#include <sys/types.h>
#include <netinet/in.h>
#include <sys/socket.h>

void* socket_worker(void* arg) {
  char buffer[MAXBUF + 1];
  char* msg = "ACK\n";
  struct sockaddr_in dest; /* socket info about the machine connecting
to us */
```

```
struct sockaddr_in serv; /* socket info about our server */
int homesocket; /* socket used to listen for incoming
connections */
socklen_t socksize = sizeof(struct sockaddr_in);

memset(&serv, 0, sizeof(serv)); /* zero the struct before
filling the fields */
serv.sin_family = AF_INET; /* set the type of connection to
TCP/IP */
serv.sin_addr.s_addr = htonl(INADDR_ANY); /* set our address to
any interface */
serv.sin_port = htons(SOCKET); /* set the server port number */

homesocket = socket(AF_INET, SOCK_STREAM, 0);

/* bind serv information to mysocket */
bind(homesocket, (struct sockaddr *) &serv, sizeof(struct
sockaddr));

/* start listening, allowing a queue of up to 1 pending
connection */
listen(homesocket, 1);
int consocket = accept(homesocket, (struct sockaddr *) &dest,
&socksize);
while (consocket) {
  send(consocket, msg, strlen(msg), 0);
  int received = -1;
  /* Receive message */
  if ((received = recv(consocket, buffer, MAXBUF, 0)) >= 0) {
    //printf("Received %s from %s \n", buffer,
    inet_ntoa(dest.sin_addr));
    snprintf(arg, sizeof(buffer), buffer);
    memset(buffer,0,sizeof(buffer));
  }
  close(consocket);
  consocket = accept(homesocket, (struct sockaddr *) &dest,
  &socksize);
}
return NULL;
}
```

Let's review the code step by step. First, you need to include the required Linux libraries in your application; `<arpa/inet.h>`, `<sys/types.h>`, `<netinet/in.h>`, and `<sys/socket.h>` are the required header files. Then, in the `socket_worker` function, you need to define the variables that are needed to create a network socket, send/receive data, and socket file descriptor. The `sockaddr_in` and `socklen_t` structs are defined in the libraries that we have included.

After variable definitions and declarations, we initialize the network socket and bind it to any interface, Wi-Fi, or Ethernet interface. Then, run a loop to accept the incoming connection and use the `recv` function to receive messages from the incoming connection, and use the `send` function to send the ACK message.

Now, copy the received message buffer to `arg`, which is the `command` variable from the main process. The main process checks the incoming message to execute the command.

Summary

In this chapter, we reviewed how we can customize the Intel Galileo Yocto Linux image by adding the home automation application. We also reviewed how we created the required files to build the application with the Yocto Project build system and install the application into the system image.

Then, we looked into the details of implementing methods to communicate with our application in Intel Galileo. We reviewed the named pipes for inter-process communication on Intel Galileo. Then we had a brief on network sockets and sample code to receive commands from defined network sockets.

In the last chapter, you try to connect to your home automation application with other platforms to manage the home automation application with remote applications. In order to do that, we will introduce you to Node.js for a basic web server to serve files from Intel Galileo and a basic web interface for remote users. Then you will follow up with a simple Android application to manage Intel Galileo applications with the network socket connection.

8
Extending Use Cases

In the previous chapters, we examined Intel Galileo and developed sample applications with C programming languages. All our applications run on the Linux operating system. We aimed to work with devices and sensors, which are frequently used in home automation systems for our applications. Finally, in the previous chapter, you had an overview of the methods to add applications into the default Linux system to act as a service in the operating system.

However, all our application user interfaces are based on the command line. The command line is a good tool to interface for developers and engineers, but it is not practical and the visualization is limited to the defined ASCII characters.

In order to make the control of the home automation application by the user or easier by switching off the wall plug anywhere you are connected, you need to check the energy consumption from any device like a smartphone, tablet, smart TV, or your game console.

We will start by investigating Node.js and will use its basic capabilities to create a basic web server. The web server will also be serving a simple web interface to send commands to the home automation application in Intel Galileo. We will follow up with a simple Android application. Our Android application will read the current status of the home from the served file and send commands from web sockets defined in the previous chapter.

This chapter will not teach Node.js and Android in depth, and so we suggest you do further reading for a quick review of JavaScript and Java programming languages. Then you need to investigate Android SDK. Let's start with Node.js in the following section.

Introducing Node.js

Node.js is a cross-platform framework to build network applications with JavaScript programming languages. Applications developed with Node.js use an event-driven method, and its most important feature is to develop a non-blocking I/O model for a network application. Node.js applications mainly use network sockets for I/O.

There are also some APIs provided to developers to interact with low-level C APIs in order to develop applications with sensors connected to platforms. Node.js is embraced by the maker community as well. They use it to build simple applications for devices like Intel Galileo. When we build the full image for an SD card, it will include Node.js and we can work with it.

The official Node.js website is `http://nodejs.org`. You can read the API documentation from the given link.

Before going further with the Node.js sample application, it is highly recommended to follow up with the following blog: `https://docs.nodejitsu.com/articles/`

The blog includes tutorials, which contain the main building blocks of our Node.js application presented in the following sections.

Using Node.js with Intel Galileo

Let's learn how to use Node.js with a "hello world" sample. We need to create a JavaScript file with a `.js` extension from the Intel Galileo terminal. You can also create the file on your host with your favorite text editor and copy the file to Intel Galileo with the SCP tool:

```
$ vi helloworld.js

console.log("Hello World");
```

We created the `helloworld.js` file. Node.js is a scripting language such as Python or shell. You just need to run it as shown here:

```
$ node helloworld.js

Hello World
```

This is the basic workaround for Node.js. We will need some basic utilities of Node.js. We will work to implement a web server, execute Linux processes, parse JSON files, serve HTML pages, and reply to incoming requests in our application.

Let's start by implementing our basic web server with Node.js. In order to create a HTTP web server with Node.js, we will include the Node.js `HTTP` module. The `HTTP` module will listen to a given socket for incoming network requests:

```
$ vi httpserver.js
var http = require('http');
http.createServer(function (request, response){
    response.writeHead({'Content-Type' : 'text/plain'});
    response.write("Hello World");
    response.end();
}).listen(2000);
```

Run this code after you save from the terminal. Then type `http://galileo_ip:2000` from any web browser you have; it will prompt `Hello World` on the web browser.

That was an introduction to Node.js. In the following section, you will learn how to develop our home server application to interact with the Smart Home native application and present a user interface to manage the home.

Developing a home automation server

In this section, we will proceed step by step on the home server application. The source code of the home automation server can be seen and downloaded from the following URL:

`https://github.com/odundar/galileo_ha/tree/master/webinterface`

Let's first import the required Node.js modules, which we need to use in the application, like the `HTTP` module.

```
    /*
     * Required NodeJS Modules to Include...
     */
    var http = require('http');
    var path = require('path');
    var fs = require('fs');
    var url = require('url');
    var exec = require('child_process').exec;
    var child;
```

We require HTTP to create a web server. We need the `path` and `fs` Node.js modules to read files on the file system. The `URL` module needs to parse the incoming requests from the client. Finally, we require the `child_process` module to execute the shell commands and write on the named pipe of the Smart Home application.

Now we will define the variables with the named pipe JSON files to read the status of devices; commands can be sent to the Smart Home application and the network port from which we want to start our web server.

```
var jsonFilePath = "/home/root/smarthome/home.json";
var commandsFilePath = "/home/root/smarthome/commands.json";
var pipePath = "/home/root/smarthome/SMARTHOMEPIPE";
var resourcePath = "res/";
//Server Listen Port
var PORT = 3000;
```

We picked `/home/root/smarthome` as our root directory to store files. We also need to change the corresponding paths inside the Smart Home application to create the named pipe file inside this directory. `res/` folder includes the images that we will show on our web interface.

Before going further in the code, we need to read the home status from a saved JSON file. This JSON file has been created by the Smart Home application. A simple saved JSON file is shown in the next few lines. It has been named as `home.json`. The Smart Home application updates this file if there is any change in the status of any device. We will read this file to get information about devices.

 JSON or JavaScript Object Notation is a format defined to make the interchange or storage of data easy. It is easier to parse than format the files as XML, and so it's getting more popular. The official JSON website is `http://www.json.org/`.

We are using the JSON file as it is very easy to parse the JSON format with Node.js and many other programming languages. You can also use XML, but it is easier with Node.js. A random sample from a `home.json` file can be seen in the following lines:

```
{
  "home": {
  "last_update_date": "Tue Jan  2 02:28:18 2015 ",
  "device": [
  {
    "id": "1",
    "name": "SHT11 Sensor",
    "status": "Active",
    "temperature": "24.000000",
    "humidity": "70.000000"
  },
  {
    "id": "2",
```

```
      "name": "Philio Multi-Sensor",
      "status": "Sleeping",
      "temperature": "76.000000",
      "lumination": "10.000000",
      "motion": "DETECTED",
      "door": "CLOSED",
      "battery": "0",
      "power_level": "Nor"
    },
    {
      "id": "3",
      "name": "Fibaro Wall Plug",
      "status": "Active",
      "switch": "ON",
      "power_meter": "0.600000",
      "energy_meter": "2.130000",
      "power_level": "Normal"
    },
.. .. .. Rest of the devices defined in this part .. .. ..
    {
      "id": "7",
      "name": "D-Link Network Camera",
      "status": "Active",
      "port": "134555744"
    } ] } }
```

Before going for the parsing method in the Node.js HTTP server application, let's make a quick revisit to the Smart Home application to check how we created this JSON file. As we mentioned in *Chapter 6*, *Home Surveillance and Extending Security Use Cases*, we defined variables in the device.h file to save an XML file. We also defined a new path for our JSON file in the device.h file as #define JSON_FILE_NAME "/home/root/smarthome/home.json".

Then we saved the latest values by using the C programming language FILE pointer. We wrote values in the form of a text file with the correct JSON formatting. The following is a piece of the update_device_json(const char* filename) function:

```
void update_device_json(const char* file_name) {
    FILE* json_file;
    json_file = fopen(file_name, "w");
    //TimeStamp for Last Update
    time_t rawtime;
    struct tm * timeinfo;
    time(&rawtime);
    timeinfo = localtime(&rawtime);
```

```
    char time_buf[BUFFER];
    sprintf(time_buf, "%s", asctime(timeinfo));
    int i = 0;
    while(time_buf[i] != '\n'){
        i++;
    }
    time_buf[i] = ' ';
    fprintf(json_file, "{\n      \"home\": {\n");
    fprintf(json_file,
    "\t\"last_update_date\": \"%s\",\n
    \t\"device\": [\n", time_buf);
    //Device 1 Temperature Sensor
    fprintf(json_file,
    "\t{\n\t\"id\": \"1\",\n
    \t\"name\": \"%s\",\n
    \t\"status\": \"%s\",\n
    \t\"temperature\": \"%f\",\n
    \t\"humidity\": \"%f\"\n\t},\n", TemperatureSensorName,
    temperature_sensor_status, temperature_c, relative_humidity);

// .. .. .. .. Some Code Here To Add Other Devices
//.. .. .. .. End of Function

    fprintf(json_file,"      \t]\n\t}\n}\n");
    fclose(json_file);
}
```

We call the `update_device_json` function when there is any change in the devices or when the thread requests updates from them. Now we can continue to our Node. js application.

The following lines are enough to read the JSON file to a JSON object named `homestatus`:

```
//Parse JSON File for Home Status
var homestatus;
fs.readFile(jsonFilePath, 'utf8', function (err, data) {
    if (err) throw err;
    homestatus = JSON.parse(data);
});
```

You can read the `last_update_date` with a simple line of code like the following one:

```
homestatus["home"]["last_update_date"]
```

If you want to read the `power_meter` value, you can use the following line:

```
homestatus["home"]["device"][2].power_meter
```

We get the second index and `power_meter` element to reach the value.

We will use the `homestatus` object to read the current status of devices. This will help us create a more dynamic user interface. As you have seen from the basic sample, you can send the text output to your browser. We will create our own HTML output when a connection request comes from an external device. In order to do a nice HTML user interface easily, with a list of devices, we will use external tools like jQuery.

 jQuery is a JavaScript library. jQuery makes it easy to manipulate HTML document styling and scripting. You can read more information about this from the jQuery official web site http://jquery.com/.

We have implemented a JavaScript function to read the JSON object and create a list with the latest status of devices. The following function takes the `response` parameter of the HTTP server. Then it creates an HTML file by writing all the corresponding lines.

```javascript
function JSONtoHtmlUI(res,message) {
  /**
   * Create HTML UI with jQuery List View
   */
  res.writeHead({'Content-Type': 'text/html'});
  res.write('<!DOCTYPE html>');
  res.write('<html>');
  res.write('<head>');
  res.write('<meta charset="UTF-8">');
  res.write('<meta name="viewport" content="width=device-width,
  initial-scale=1">');
  res.write('<link rel="stylesheet" href=
  "http://code.jquery.com/mobile/1.4.5/jquery.mobile-
  1.4.5.min.css">');
  res.write('<script src=
  "http://code.jquery.com/jquery-1.11.1.min.js"></script>');
  res.write('<script src=
  "http://code.jquery.com/mobile/1.4.5/jquery.mobile-
  1.4.5.min.js"></script>');
  res.write('</head>');
  res.write('<body>');
  res.write('<div data-role="page" id="pageone">');
```

```
    res.write('<div data-role="main" class="ui-content">');
    if(message != ""){
       res.write('<p> Message: ' + message + '</p>'); }
    res.write('<h4> Home Status: ' +
    homestatus["home"]["last_update_date"] + '</h4>');
    res.write('<ul data-role="listview" data-inset="true">');
    // List Devices
    // SHT11 Sensor
    res.write('<li>');
    res.write('<a href=/>');
    res.write('<img src="'+ resourcePath + 'sht11.png">');
    res.write('<h3>' + homestatus["home"]["device"][0].name +
    '</h3>');
    /** In this part of this function we are repeating similar
    responses for other devices **/
    res.write('</ul>');
    res.write('</div>');
    res.write('</div>');
    res.write('</body>');
    res.write('</html>');
    res.end();
}
```

Information about each device has been added inside the `<li>` tag to create a row in the list. For each device, we loaded the device's image to the list row and added the corresponding text. In the first few lines, we defined the jQuery links.

We will respond with the above function to show the user interface when the user makes a connection to our application. In case the user makes a request to switch off the lamp holder or wall plug, we need to handle the request and send a command to the Smart Home application. We need to define another function to send commands to the Smart Home application pipe. We can send commands to the Smart Home's pipe from the Linux shell. Node.js includes the `child_process` module and it can execute shell commands. Therefore, we used the `echo` command from the Linux shell to write the corresponding command to the Smart Home application, as shown here:

```
function sendCommand(command_id) {
    // Parse Commands JSON
    var commands = JSON.parse(fs.readFileSync(commandsFilePath,
    'utf8'));
    console.log("Parsing Commands JSON...");
    var command = 'echo ' +
    commands["commands"]["command"][command_id].text + ' > ' +
    pipePath;
    console.log("Executing....:" + command);
```

```
    child = exec(command, function(error, stdout, stderr) {
      if (error != null) {
        console.log('exec error: ' + error);
      }
    });
  }
```

The `sendCommand` function first reads the `commands.json` file from the given path to a JSON object. Then it executes the command according to the given command ID. An example of this is `echo   switch on 4 > /home/root/smarthome/` `SMARTHOMEPIPE`.

The `commands.json` file has been created with the reference of the defined commands in the Smart Home application; the file is static and doesn't change at runtime. The following lines form a part of the `commands.json` file:

```
{
  "commands": {
    "command": [
      {
        "id": "1",
        "text": "show home status"
      },
.. .. .. .. .. .. Other Commands .. .. .. .. .. ..
      {
        "id": "5",
        "text": "switch on 3"
      },
.. .. .. .. .. .. Other Commands .. .. .. .. .. ..
      {
        "id": "11",
        "text": "help"
      } ] } }
```

We are done with helper functions and variables. Now we will define our HTTP server's `handler` function in the following lines to define how to respond to the incoming request:

```
function handler(request, response) {
  if (request.method !== 'GET') {
    response.writeHead(405);
    response.end('Unsupported request method', 'utf8');
    return;
  }
  console.log("Parsing Device File...")
```

```
fs.readFile(jsonFilePath, 'utf8', function (err, data) {
  if (err) throw err;
  homestatus = JSON.parse(data);
});
var request = url.parse(request.url, true);
var action = request.pathname;
// Consolo Output for Request
console.log(action);
/**
* Handle HTTP Get Requests
*/
if (action === '/'){
  JSONtoHtmlUI(response,"");
  return;
} else if(action == '/switch4'){
  console.log("Switch Request....");
  var switch_action;
  // Switch off
  if(homestatus["home"]["device"][3].switch == "ON"){
    switch_action = "OFF";
    sendCommand(7);
  } else { // Switch on wall plug
    switch_action = "ON";
    sendCommand(6)
  }
  homestatus["home"]["device"][3].switch = switch_action;
  JSONtoHtmlUI(response,"Switched " +
  homestatus["home"]["device"][3].name + " " + switch_action);
  return this;
  // Switch On or Off LampHolder
} else if(action == '/switch3'){
  console.log("Switch Request.....");
  var switch_action;
  if(homestatus["home"]["device"][2].switch == "ON"){
    switch_action = "OFF";
    sendCommand(5);
  } else{
    switch_action = "ON";
    sendCommand(4);
  }
  homestatus["home"]["device"][2].switch = switch_action;
  JSONtoHtmlUI(response,"Switched " +
  homestatus["home"]["device"][2].name + " " + switch_action);
  return this;
```

```
    // Capture Frame from Network Camera
  } else if(action == '/capture'){
    sendCommand(8);
    JSONtoHtmlUI(response,"Captured From Network Camera");
    return this;
  }
  /**
   * Serve Requested Static Files
   */
  var filePath = path.join(__dirname, action);
  var stat = fs.statSync(filePath);
  fs.exists(filePath, function (exists) {
    if (!exists) {
      // 404 missing files
      response.writeHead(404, {'Content-Type': 'text/plain' });
      response.end('404 Not Found');
      return;
    }
    var readStream = fs.createReadStream(filePath);
    readStream.on('data',function(data){
    response.write(data);
    });
    readStream.on('end',function(data){
    response.end();
    });
  });
  return;
}
```

Our handler function parses the incoming requests from the browser. It takes the appropriate action for each request. If a file requested is `http://galileo_ip:3000/home.json`, it servers the corresponding file. If the user requests a resource file with the URL `http://galileo_ip:3000/res/sht11.png`; the browser loads the image. If the user wants to switch the wall plug on or off, the user requests this with `http://galileo_ip:3000/switch3`.

Finally, we create the HTTP server with only one line of code, as shown in the following line:

```
http.createServer(handler).listen(PORT);
```

We have completed our home server application and are now ready to run the application. Save the code to a JavaScript file. In this sample, we saved the code to the `homeserver.js` file inside the `/home/root/smarthome` folder, which is shown in the following lines:

```
$ cd /home/root/smarthome
$ node homeserver.js
```

Now our Node.js application is working. Let's first try connecting from our host PC browser; type the Galileo IP address and port, which is `http://192.168.2.235:3000` in our case. The user interface has been tested only on the Firefox and Chrome browsers on a PC. Look at the following screenshot from a desktop Firefox browser on the left and an Android mobile browser on the right:

Now we can click on the **Fibaro Wall Plug** item on the list to switch it on or off from our smartphone.

You can improve the application by adding more features and beautify the user interface for a better experience. If your devices are different to the ones presented here, you can change and add custom requests to handle using the Node.js application.

We will keep the Node.js application running to serve the `home.json` and `commands.json` files for the next section's smartphone application.

Use cases with Node.js

In the previous chapter, we made an application to run as a web server, present a web interface and send commands to the Smart Home application. However, you can extend your applications with Node.js using some very cool options. As it is easy to program network applications with Node.js, you can connect any cloud or web service to Node.js.

One idea could be to integrate your Node.js application with Twitter. Twitter provides APIs for many platforms; there are many open source APIs for Twitter as well and they can be used easily with Node.js.

It is a very nice option to use the tweet feature if there is any motion detected at home. You can tweet if there is any change in the devices such as a flood or gas leak detected. We have sent commands from a web interface; it is also possible to define your application with the help of the Twitter API, to read the incoming direct messages to send commands to the Smart Home application. For example, you can create a new account for the home automation application. A Node.js application can send tweets if there is any change at home and can read Twitter direct messages to send commands to the Smart Home application.

Twitter is only one example; there are many free services with public APIs. You can add the feature to your home to control it wherever you are connected.

In this section, we tried to cover the basics of Node.js with a sample application. There are tons of use cases that you can add to Node.js to polish the home automation application. Now we will proceed by developing an Android application.

Introducing Android

We have developed a Node.js application as a web server as well as to present an HTML user interface in order for users to interact with the home automation application. Another way to interact with the Smart Home application on Intel Galileo could be by developing native applications for your device; the device could be your personal computer, tablet, game console, smart TV, tablet, or smartphone.

The native application can send commands to the Smart Home application network socket, which we introduced in the previous section. You can also simply interact with the Node.js application from your native application instead of using a web browser. As Android is very popular, we will show how to develop a native application on our Android smartphone to interact with the Smart Home application.

If you are not used to Android, this section may be hard to grasp. It is recommended to review the Android operating system, SDK, and the application development environment.

Developing Android applications with Android SDK requires you to know the Java programming language. It is also a good idea to review the Java programming language.

The following URL's include tutorials and examples for Android application development. It would be very useful for you to follow them: `http://developer.android.com/index.html`, `http://developer.android.com/training/index.html`, `http://www.vogella.com/tutorials/AndroidListView/article.html`

Before starting to develop with Android SDK, you need to install Java onto your host PC. Then you should install Android SDK; an easy way to start is to download the Android ADT Bundle, which includes both Eclipse IDE and Android SDK. Another option for this is to download Android Studio. You can find more information at `http://developer.android.com/sdk/index.html`. We will be using Eclipse IDE to develop our Android application.

Developing an Android application for home automation

In this section, we will describe the main steps to develop a home automation application for Android. You can see and download the source code of the application from the repository at `https://github.com/odundar/galileo_ha/tree/master/smarthome_app`.

In order to start the development, we need to create the Android application project from IDE. In our sample, we named the project `SmartHome`. While creating the application, it prompts you to create an Android activity. For this application, we will have one Android activity named `SmartHome` in the `SmartHome.java` file. The Android activity handles the entire user interface, workload, and logic. When we create the activity while creating the Android project, it also populates an XML file, which holds the Android application's user interface data and elements. It is automatically named as `activity_smart_home.xml`.

Look at the following screenshot from our Android project. You can see the Java classes and XML files for this project. We also copied all the images of the devices to the `res/drawable-hdpi` folder under the `project` folder. In the Node.js application, we are loading the images from Intel Galileo but, to make the application faster, it is better to use local resources in the Android application.

As you can see in the project folder, we also have the **DeviceItem.java**, **HomeDeviceAdapter.java** classes, and the **row_layout.xml** file. We created these files to build a custom list to store the image of the device, the name of the device and the latest status of the device and show them in an Android `ListView` widget. Android `ListView` widgets can be assigned to a data adapter and automatically fill the list with the given items.

Let's first check our `ListView` widget's custom row layout from the `row_layout.xml` file. XML files store the user interface structure. In order to show an image in an Android device, the `ImageView` class can be used; to write text on an Android application, `TextView` objects are used.

```
<LinearLayout xmlns:android="http://schemas.android.com/apk/res/
android"
   android:layout_width="fill_parent"
   android:layout_height="fill_parent"
   android:orientation="vertical" >
   <ImageView
     android:id="@+id/photo"
<!--Here the Properties of View described-->/>
     <TextView
        android:id="@+id/name"
<!--Here the Properties of View described--> />
     <TextView
<!--Here the Properties of View described--> />
   </LinearLayout>
```

The `DeviceItem` class is a Java class to store device information to fill each row.

```
public class DeviceItem {
   private String deviceName;
   private String deviceStatus;
   private int deviceImage;
   /*
   * Rest of the code is getters and setters of private members of
   this function
   */
   public String getDeviceName() {
     return deviceName;
   }

   public void setDeviceName(String deviceName) {
     this.deviceName = deviceName;
   }
//// Some Code Here, deviceStatus and deviceImage Getters and
Setter////
   }
```

The `HomeDeviceAdapter` is a child class, which inherits the `BaseAdapter` class from Android SDK; it is required to override its `getView` function to fill the `ListView` widget with the given items. In our case, it will pass the `DeviceItem` objects to fill the `ListView` widget of our application to show the home status.

```java
public class HomeDeviceAdapter extends BaseAdapter {
  private final ArrayList<DeviceItem> itemsArrayList;
  private LayoutInflater inflator;
  public HomeDeviceAdapter(Context context,
  ArrayList<DeviceItem> itemsArrayList) {
    super();
    this.itemsArrayList = itemsArrayList;
    inflator = LayoutInflater.from(context);
  }
  /* … Here Other Inherited Methods from BaseAdapter  Methods Code
  …. */
  @Override
  public View getView(int position, View convertView, ViewGroup
  parent) {
    ViewHolder holder;
    if (convertView == null) {
      convertView = inflator.inflate(R.layout.row_layout, null);
      holder = new ViewHolder();
      holder.deviceName = (TextView)
      convertView.findViewById(R.id.name);
      holder.deviceStatus = (TextView) convertView
      .findViewById(R.id.itemDescription);
      holder.deviceImage = (ImageView) convertView
      .findViewById(R.id.photo);
      convertView.setTag(holder);
    } else {
      holder = (ViewHolder) convertView.getTag();
    }
    holder.deviceName.setText(itemsArrayList.get(position).
    getDeviceName());
    holder.deviceStatus.setText(itemsArrayList.get(position)
    .getDeviceStatus());
    holder.deviceImage.setImageResource(itemsArrayList.get
    (position).getDeviceImage());
    return convertView;
  }
  static class ViewHolder {
    TextView deviceName;
    TextView deviceStatus;
    ImageView deviceImage;
  }
}
```

In the preceding code, we also defined the private members itemsArrayList parameter and inflator to use in the getView method. The LayoutInflator inflator needs to be created from the current context of the application to populate rows of the ListView widget in the current application. The itemsArrayList parameter contains the passed DeviceItem objects.

Now we will proceed to implement the SmartHome activity to do all the work. When we first create the activity during project creation, it inherits the Android SDK, Activity objects' properties, and the onCreate(Bundle savedInstanceState) function. The onCreate function is called in Android applications when you run the application for the first time; it is mostly used to initialize the user interface elements or initialize the required variables to be used during the application. Let's first add the required variables to use in the activity class:

```
private URL homeJSONUrl;
  private URL commandsJSONUrl;
  private String lastUpdateDate;
  private JSONObject homeInfo;
  private String[] commands;
  private String[] deviceNames;
  private String[] deviceStatus;
  private Integer[] deviceImages = { R.drawable.sht11,
    R.drawable.philio,
    R.drawable.fibaro, R.drawable.everspring, R.drawable.flood,
    R.drawable.mq9, R.drawable.dlink };
  private Socket galileSocket;
  private static final int GALILEOPORT = 3500;
  private static final String GALILEO_IP = "192.168.2.235";
  volatile public boolean filesdownloaded = false;
  volatile public boolean socketcreated = false;
  ArrayList<DeviceItem> listItems;
  ListView deviceListView;
  TextView updateDateText;
```

We have defined the socket port, the Intel Galileo IP address, the user interface elements to manipulate them in the code, and arrays to store information read from JSON files.

We need to handle network operations in separate threads. The Android activity is a thread needed to handle user interaction and user interface elements drawing to a smartphone screen, and so it doesn't allow us to interrupt the thread to fetch information from the network. We will define three different threads to handle JSON file readings and the network socket creation.

The following code shows the threads that we will use in the code:

```
Thread socketThread = new Thread(new Runnable() {
  @Override
  public void run() {
    // TODO Auto-generated method stub
    InetAddress serverAddr;
    try {
      serverAddr = InetAddress.getByName(GALILEO_IP);
      galileSocket = new Socket(serverAddr, GALILEOPORT);
      socketcreated = true;
    } catch (UnknownHostException e) {
      e.printStackTrace();
    } catch (IOException e) {
      e.printStackTrace();
    }
  }
});
/**
 * JSON Worker to Fetch JSON Files from Intel Galileo
 */
Thread jsonFetcher = new Thread(new Runnable() {
  @Override
  public void run() {
    // TODO Auto-generated method stub
    try {
      // Initialize URLs
      homeJSONUrl = new
      URL("http://192.168.2.235:3000/home.json");
      commandsJSONUrl = new URL(
        "http://192.168.2.235:3000/commands.json");
      getHomeStatus();
      getCommandsList();
      // Fill String Arrays
      initalizeHomeArray();
      fillListAdapter();
      filesdownloaded = true;
    }
    /* Catch Block Code */
  }
});
/**
 * Update Adapter
```

```
     * Periodically Read the home.json File and Update ListView
     Adapter
     */
     Thread updater = new Thread(new Runnable() {
       @Override
       public void run() {
         // TODO Auto-generated method stub
         while (true) {
           try {
             try {
               Thread.sleep(10000);
             } catch (InterruptedException e) {
               e.printStackTrace();
             }
             // Initialize URLs
             homeJSONUrl = new
             URL("http://192.168.2.235:3000/home.json");
             getHomeStatus();
             // Fill String Arrays
             updateHomeArray();
             // Clear Adapter
             listItems.clear();
             fillListAdapter();
           }
  /* Catch Block Code */
           }
         }
       });
```

The `updater` thread works in ten second periods to read the `home.json` file periodically to update the `ListView` widget if there is any change. The `jsonFetcher` parameter works at the start of the application to fetch the initial states of JSON files. The `socketThread` parameter creates the network socket which we will use to send commands to the Smart Home application on Intel Galileo.

Let's proceed to our `onCreate` function to see how we initialize the user interface for the application. In the application, we first set the application layout, which is `activity_smart_home.xml`. In the main user interface layout, we have the header text, a list of devices, and finally an `update` button to update the list any time we want. The following lines show the XML file content:

```
<LinearLayout
xmlns:android="http://schemas.android.com/apk/res/android"
  xmlns:tools="http://schemas.android.com/tools"
  android:layout_width="fill_parent"
```

```
android:layout_height="fill_parent"
android:background="#bdc3c7"
android:orientation="vertical"
tools:context="com.galileha.smarthome.SmartHome" >
<TextView
  android:id="@+id/skip"
  <!--Here the Properties of View described--> />
<TextView
  android:id="@+id/updateDate"
  <!--Here the Properties of View described--> />
  <LinearLayout
    android:layout_width="fill_parent"
    android:layout_height="wrap_content" >
  <ListView
    android:id="@+id/list"
    android:layout_width="fill_parent"
    android:layout_height="380dp" >
  </ListView>
  </LinearLayout>
  <Button
    android:id="@+id/update"
    <!--Here the Properties of View described--> />
</LinearLayout>
```

Then we initialize the user interface elements and start threads to fetch the JSON files. The following function is used to fetch the home.json file inside the Android application:

```
public void getHomeStatus() throws IOException,
MalformedURLException, JSONException {
  // Set URL
  // Connect to Intel Galileo get Device Status
  HttpURLConnection httpCon = (HttpURLConnection) homeJSONUrl
  .openConnection();
  httpCon.setReadTimeout(10000);
  httpCon.setConnectTimeout(15000);
  httpCon.setRequestMethod("GET");
  httpCon.setDoInput(true);
  httpCon.connect();
  // Read JSON File as InputStream
  InputStream readStream = httpCon.getInputStream();
  Scanner scan = new Scanner(readStream).useDelimiter("\\A");
  // Set stream to String
  String jsonFile = scan.hasNext() ? scan.next() : "";
  // Initialize serveFile as read string
```

```
      homeInfo = new JSONObject(jsonFile);
      httpCon.disconnect();
   }
```

We finally filled our adapter to populate the ListView widget with corresponding data in the onCreate function. Then we defined a listener, which is responsible for sending a command according to the clicked list item.

```
@Override
   protected void onCreate(Bundle savedInstanceState) {
      super.onCreate(savedInstanceState);
      setContentView(R.layout.activity_smart_home);
      deviceListView = (ListView) findViewById(R.id.list);
      updateDateText = (TextView) findViewById(R.id.updateDate);
      listItems = new ArrayList<DeviceItem>();
      jsonFetcher.start();
      while (!filesdownloaded);
      HomeDeviceAdapter deviceAdapter = new HomeDeviceAdapter(this,
      listItems);
      deviceListView.setAdapter(deviceAdapter);
      deviceListView.setOnItemClickListener(new
      OnItemClickListener() {
        @Override
        public void onItemClick(AdapterView<?> parent, View view,
        int position, long id) {
          JSONObject homeDevices;
          JSONArray devices;
          JSONObject clickedObject = null;
          try {
            homeDevices = homeInfo.getJSONObject("home");
            devices = (JSONArray)
            homeDevices.getJSONArray("device");
            clickedObject = devices.getJSONObject(position);
          } catch (JSONException e) {
            // TODO Auto-generated catch block
            e.printStackTrace();
          }
          // Switch ON/OFF Wall Plug or LampHolder
  /* We check for other positions in this code block */
          } else {
            Toast.makeText(SmartHome.this, "No Available Command for
            Selected Device",   Toast.LENGTH_LONG).show();
          }
        }
      });
```

```
        updateDateText.setText(lastUpdateDate);
        socketThread.start();
        while (!socketcreated);
        updater.start();
        Log.d("SOCKET", "Socket Thread Started");
    }
```

When we click on a list item, it sends the corresponding command to the Smart Home application network socket. The following methods get a string parameter to send to the network socket. The use of the function can be seen in the onCreate function. We also introduced the update button, which updates the user interface elements when we click on it. You can define which function to call when the button is clicked in the XML layout. In this case, we defined the android:onClick="onUpdate" method. You can see the onUpdate function shown here:

```
public void writeToSocket(String message) {
    try {
        PrintWriter out = new PrintWriter(new BufferedWriter(
            new OutputStreamWriter(galileSocket.getOutputStream())),
              true);
        out.println(message);
      Log.d("SOCKET", "Message " + message + " Sent to Socket");
    } catch (IOException e) {
        // TODO Auto-generated catch block
        e.printStackTrace();
    }
}
public void onUpdate(View v) {
    updateDateText.invalidate();
    deviceListView.invalidateViews();
    deviceListView.refreshDrawableState();
    updateDateText.setText(lastUpdateDate);
    updateDateText.refreshDrawableState();
    Log.d("JSON", "Updated Views....");
}
```

Finally, our application will look like the following image: a screenshot from the Android device, and we'll be able to see the devices' status and manage them.

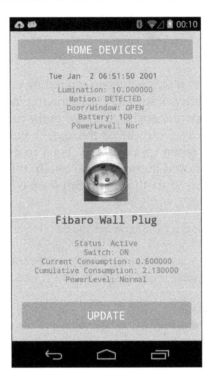

There's more in Android

Android is very rich in supporting libraries to enhance your application with cool features.

It is also possible with Android to directly connect to network camera and get a stream to your application. Android includes Media API to play the network video. More information about Media API can be found from `http://developer.android.com/reference/android/media/package-summary.html`.

You can use Bluetooth API to communicate with Intel Galileo from the Android application. You would need a Bluetooth device attached to Intel Galileo and need to program it to send the right information to the smartphone when it is connected. For more information about the Android Bluetooth API, you can access the following link: `http://developer.android.com/reference/android/bluetooth/package-summary.html`

These are some cool features that can be integrated with the application; you can delve into Android API for more on this. In this book, we only go over the Android operating system, but you may also try creating a similar application for iPhone and Windows Phone according to your experience and interest.

Adding voice control to home automation

Nowadays it is popular to use voice-directed personal assistants such as Apple's Siri, Google's Now and Microsoft's Cortana. Adding speech or sound recognition is a nice addition to your home automation system.

Speech and voice recognition libraries exist for almost all platforms. Depending on your design, you can add voice control to home automation. Let's start with Android.

Voice control with Android

An easy way to control your application with voice commands is to extend the Android sample application with the Android Speech Recognizer library.

Information about the Android Speech Recognizer library can be found at `http://developer.android.com/reference/android/speech/SpeechRecognizer.html`.

For our sample application, we would create a new thread to work as a speech listener by using the `SpeechRecognizer` library and listening to the incoming voice.

The `SpeechRecognizer` library populates text, and so you can compare if any of them matches the commands we loaded from the JSON file. If it does, we can send a command to the Smart Home application. You may also define better sentences to match the commands and send the corresponding commands to Intel Galileo. For example, you can define a phrase such as `Open Lights` and this text will send the command `switch on 4`, which turns on the lamp holder.

Voice recognition with Intel Galileo

The easiest way to add a microphone to get sound from the environment with Intel Galileo is to use a USB microphone. It is also possible for you to connect an analog audio sensor to get raw audio to Intel Galileo. It will take more time to handle raw audio if you do not have much experience with audio.

After you have added a microphone or sound device, you need to add **ALSA** drivers to get audio from the hardware device and sound from the environment. ALSA drivers and the Linux library can be built with the Yocto Project by following the steps that were described in the previous chapters.

When we get the audio, we need to use the audio library to read sound from the ALSA driver and feed the speech recognizer. A suggested open source library for audio handling is **PortAudio**. For speech recognition, **Sphinx** is one of the most popular speech recognition libraries to use in the Linux environment. **Pocketsphinx**, the core of Sphinx, provides a C API to access and enable your platform for speech recognition.

More information about libraries can be found from the following links:

- ALSA: `http://www.alsa-project.org/main/index.php/Main_Page`
- PortAudio: `http://www.portaudio.com/`
- Sphinx: `http://cmusphinx.sourceforge.net/`

Summary

Here we come to the end of the book. We tried to develop a piece of home automation with various devices and technologies to help you understand the home automation concept, technologies, Intel Galileo, the Linux operating system, the Yocto Project, and so on. We also presented new technologies to improve the home automation system with better user interaction methods.

We used Node.js to create an application to run as a web server and to create a basic HTML user interface and communicate with native applications to send commands to it via web browser actions.

Then we showed how to develop an Android application to read the devices status from Intel Galileo and send commands to the Smart Home application from the network socket. We were not able to study Android too deeply, but it would be a good move for you to learn more about Android.

Index

I

image
  capturing, from camera 111, 112
Insteon
  about 30, 31
  URL 31
Insteon USB Interface
  URL 58
Intel Galileo
  about 2, 3
  booting 7, 8
  cameras, using with 107
  connecting 7, 8
  connecting to local network, via Telnet 9
  development environment, setting up
      for 17, 18
  firmware, upgrading on 16
  gas sensors, using with 87-91
  hardware specifications 4
  image, capturing from camera 111, 112
  kernel modules, building for 57
  Linux filesystems, building for 11-14
  Node.js, using with 136, 137
  OpenCV, building for 107, 108
  Philips Hue, using with 67
  remote wall plug, controlling from 60-62
  SD card, preparing to boot 14, 15
  security sensors 85
  sensor based applications,
      developing with 42
  sensor, connecting to 44
  software specifications 5, 6
  status of remote devices, reading from 80
  URL, for community 3, 10
  URL, for downloading kernel source code 3
  used, for designing home automation
      project 37-39
  using, for home automation 3, 4
  V4L2, building for 107, 108
  video, saving from camera 113-115
  video, streaming from 115
  Z-Wave messages, handling from 75-77
Intel Galileo Gen 1
  reference link 2
Intel Galileo Gen 1 USB host
  about 2
  URL, for example cable 2

Intel Galileo IO Mapping document
  URL 46
Intel Galileo SDK
  building 16, 17
Intel Quark board support package
  URL, for downloading 10

J

jQuery
  URL 141
JSON
  URL 138

K

kernel modules
  building, for Intel Galileo 57
KNX 30

L

light intensity 65
light sensors
  using 65, 66
Linux
  customizing, with Yocto project 125, 126
  development environment,
      setting up for 18
Linux filesystems
  building, for Intel Galileo 11-14
Linux image, for Intel Galileo
  building, with Yocto Project 10, 11
Linux Media Center Edition (Linux MCE)
  about 33
  URL 33
Linux Terminal
  applications, building on 19
LM35 42
Lux 65

M

magnetic sensors
  about 86, 87
  reference link 87
message parsing system
  wrapping up 98-102

Thank you for buying
Home Automation with Intel Galileo

About Packt Publishing

Packt, pronounced 'packed', published its first book, *Mastering phpMyAdmin for Effective MySQL Management*, in April 2004, and subsequently continued to specialize in publishing highly focused books on specific technologies and solutions.

Our books and publications share the experiences of your fellow IT professionals in adapting and customizing today's systems, applications, and frameworks. Our solution-based books give you the knowledge and power to customize the software and technologies you're using to get the job done. Packt books are more specific and less general than the IT books you have seen in the past. Our unique business model allows us to bring you more focused information, giving you more of what you need to know, and less of what you don't.

Packt is a modern yet unique publishing company that focuses on producing quality, cutting-edge books for communities of developers, administrators, and newbies alike. For more information, please visit our website at www.packtpub.com.

About Packt Open Source

In 2010, Packt launched two new brands, Packt Open Source and Packt Enterprise, in order to continue its focus on specialization. This book is part of the Packt Open Source brand, home to books published on software built around open source licenses, and offering information to anybody from advanced developers to budding web designers. The Open Source brand also runs Packt's Open Source Royalty Scheme, by which Packt gives a royalty to each open source project about whose software a book is sold.

Writing for Packt

We welcome all inquiries from people who are interested in authoring. Book proposals should be sent to author@packtpub.com. If your book idea is still at an early stage and you would like to discuss it first before writing a formal book proposal, then please contact us; one of our commissioning editors will get in touch with you.

We're not just looking for published authors; if you have strong technical skills but no writing experience, our experienced editors can help you develop a writing career, or simply get some additional reward for your expertise.

Raspberry Pi Home Automation with Arduino

ISBN: 978-1-84969-586-2   Paperback: 176 pages

Automate your home with a set of exciting projects for the Raspberry Pi!

1. Learn how to dynamically adjust your living environment with detailed step-by-step examples.

2. Discover how you can utilize the combined power of the Raspberry Pi and Arduino for your own projects.

3. Revolutionize the way you interact with your home on a daily basis.

Arduino Home Automation Projects

ISBN: 978-1-78398-606-4   Paperback: 132 pages

Automate your home using the powerful Arduino platform

1. Interface home automation components with Arduino.

2. Automate your projects to communicate wirelessly using XBee, Bluetooth, and Wi-Fi.

3. Build seven exciting, instruction-based home automation projects with Arduino in no time.

Please check **www.PacktPub.com** for information on our titles

Arduino Robotic Projects

ISBN: 978-1-78398-982-9   Paperback: 240 pages

Build awesome and complex robots with the power of Arduino

1. Develop a series of exciting robots that can sail, go under water, and fly.

2. Simple, easy-to-understand instructions to program Arduino.

3. Effectively control the movements of all types of motors using Arduino.

4. Use sensors, GSP, and a magnetic compass to give your robot direction and make it lifelike.

BeagleBone Home Automation

ISBN: 978-1-78328-573-0   Paperback: 178 pages

Live your sophisticated dream with home automation using BeagleBone

1. Practical approach to home automation using BeagleBone; starting from the very basics of GPIO control and progressing up to building a complete home automation solution.

2. Covers the operating principles of a range of useful environment sensors, including their programming and integration to the server application.

3. Easy-to-follow approach with electronics schematics, wiring diagrams, and controller code all broken down into manageable and easy-to-understand sections.

Please check **www.PacktPub.com** for information on our titles

www.ingramcontent.com/pod-product-compliance
Lightning Source LLC
Chambersburg PA
CBHW082119070326
40690CB00049B/3850